Lectures on
Computational Fluid Dynamics, Mathematical Physics, and Linear Algebra

Lectures on
Computational Fluid Dynamics, Mathematical Physics, and Linear Algebra

Karl Gustafson
Universary of Colorado, Boulder

Editors

T Abe
Shibaura Institute of Technology
K Kuwahara
The Institute of Space and Astronautical Science

World Scientific
Singapore • New Jersey • London • Hong Kong

Published by

World Scientific Publishing Co. Pte. Ltd.

P O Box 128, Farrer Road, Singapore 912805

USA office: Suite 1B, 1060 Main Street, River Edge, NJ 07661

UK office: 57 Shelton Street, Covent Garden, London WC2H 9HE

British Library Cataloguing-in-Publication Data
A catalogue record for this book is available from the British Library.

LECTURES ON COMPUTATIONAL FLUID DYNAMICS, MATHEMATICAL PHYSICS, AND LINEAR ALGEBRA

ISBN 981-02-3213-6

Printed in Singapore by Uto-Print

Foreword

In recent years there has been remarkable progress in mathematical science,including mathematics, mathematical physics and other applied mathematics,as a basis for every technology. It was a significant event for those who had interests in the field above and for the field itself in Japan that we had enjoyed Professor Karl Gustafson's lectures, by inviting him who energetically carried on surprisingly broad research in this fundamental scientific field through the present, and was one of scientists with the worldwide remarkable contributions to the field as a mathematician, to Japan last year. Particularly, it seemed to be promising that young people agressively attended the lectures, mingled with domestic and external celebrated scientists.

Now, this book is organized as follows, in accordance with a series of six lectures (see p.3 for their details) done by Professor Gustafson in Tokyo and Kyoto from October 26 through November 2,1995 :

Part I.   Recent Developments in Computational Fluid Dynamics,

Part II.   Recent Developments in Mathematical Physics,

Part III.   Recent Developments in Linear Algebra,

and

Four Comments on the Lectures above.

Each Part forms the core of the book, and its outline is described in *perspective* of each Part. We would like to advise readers to look at it. And, we hope that the comments added to the lectures are interesting and useful for readers.

Although the lectures held during the restricted short term are only part of his numerous research themes, as easily seen, these are the subjects selected in deep connection with the latest problems in mathematics, physics and engineering, and give the most new informations including their latest results and unpublished ones. In spite of it,the book plays the role of a guide readable for further advanced studies fit for the wide reading public from beginners to experts in each field, which the editors strongly desired. This is shown by the following features :

(i) The brief and attentive *perspectives* for the contents clarify the backgrounds and the meanings of the themes.

(ii) Essential concepts,various methods and concrete techniques for the themes are efficiently given without a dogmatic description style.

(iii) The discussions are not restricted within themselves, and their contents , and critical minds extended for related subjects are synthetically presented in a well-balanced style.

(iv) On the whole the book suggests stimulating problems and subjects to be solved near future, and in some cases gives conjectures.

(v) The attentive Comments and Bibliographies, and a large number of References play excellent roles of guidance for ambitious researchers .

Although it is needless to say that considerable effort is required for thorough understanding of the contents, and there may be a few readers discontented with insufficiency of mathematical rigor including proofs on account of the limited number of pages for the book,it is pleasant to survey at least the present conditions of the special fields.

We express a high regard for the author's strong enthusiasm and power which he gave a ready consent to our request of writing the book, and has completed this excellent one, despite the hard work after returning to his country.

We would like to recommend this book by the great scholar supported by the various research activities and the remarkable achievements in these, a broad and high intelligence, and a wealth of humanity, to the people concerned in the fields.

Takehisa Abe
Shibaura Institute of Technology

Kunio Kuwahara
The Institute of Space and Astronautical Science

March 1996
Tokyo

Acknowledgments

I wish to express deeply my gratitude to Professor Huzihiro Araki who gave me variously careful consideration and warm help throughout the event as one of the organizers. And at the same time, I sincerely thank Professors Kunio Kuwahara, Mitsuharu Ohtani, Hisashi Okamoto and Miki Wadachi for their good offices and friendship on the occasion of holding the lectures and of putting them into practice as the organizers, also I express my thanks to all the participants for their livening up of the meetings. And I am grateful to Professors K.Kuwahara and Shuichi Tasaki for their pleasant cooperation with me, as the promoters for the event. Further I would like to express sincere thanks to the enterprises Kyoritsu Publisher, Nihon Process, Nomura Research Institute, Quick Electronic Information and others, including Kaigai Publications, Ltd. for their warm understanding and financial helps. In addition to the above, I am greatly indebted to various many people concerned with the event for their kind helps referring to public and private matters.

On the occasion of the publication of the book, I would like to thank the contributors of the comments, Professors K.Kuwahara, M.Ohtani and ,in particular, Izumi Ojima, for their pleasant cooperation for my requests. And Professor M.Wadachi suggested me a useful advice on the publication. Furthermore, I thank heartily Mrs. Elizabeth Stimmel for her devotion to the work of wonderful typing of the manuscript, and also my thanks are expressed to those who spared no pains for the publication. Finally, I wish to heartily thank the president Hajime Kuroda and the director Tomiji Ohno of Kaigai Publications who carried out the plan to publish the book for their deep understanding and great help, and my thanks are given to their editorial staff as well.

Takehisa Abe

April 1996
Tokyo

Prof.Gustafson, T.Abe and his wife at Asakusa(the old part of Tokyo), in pauses of the lectures (October 29, 1995)

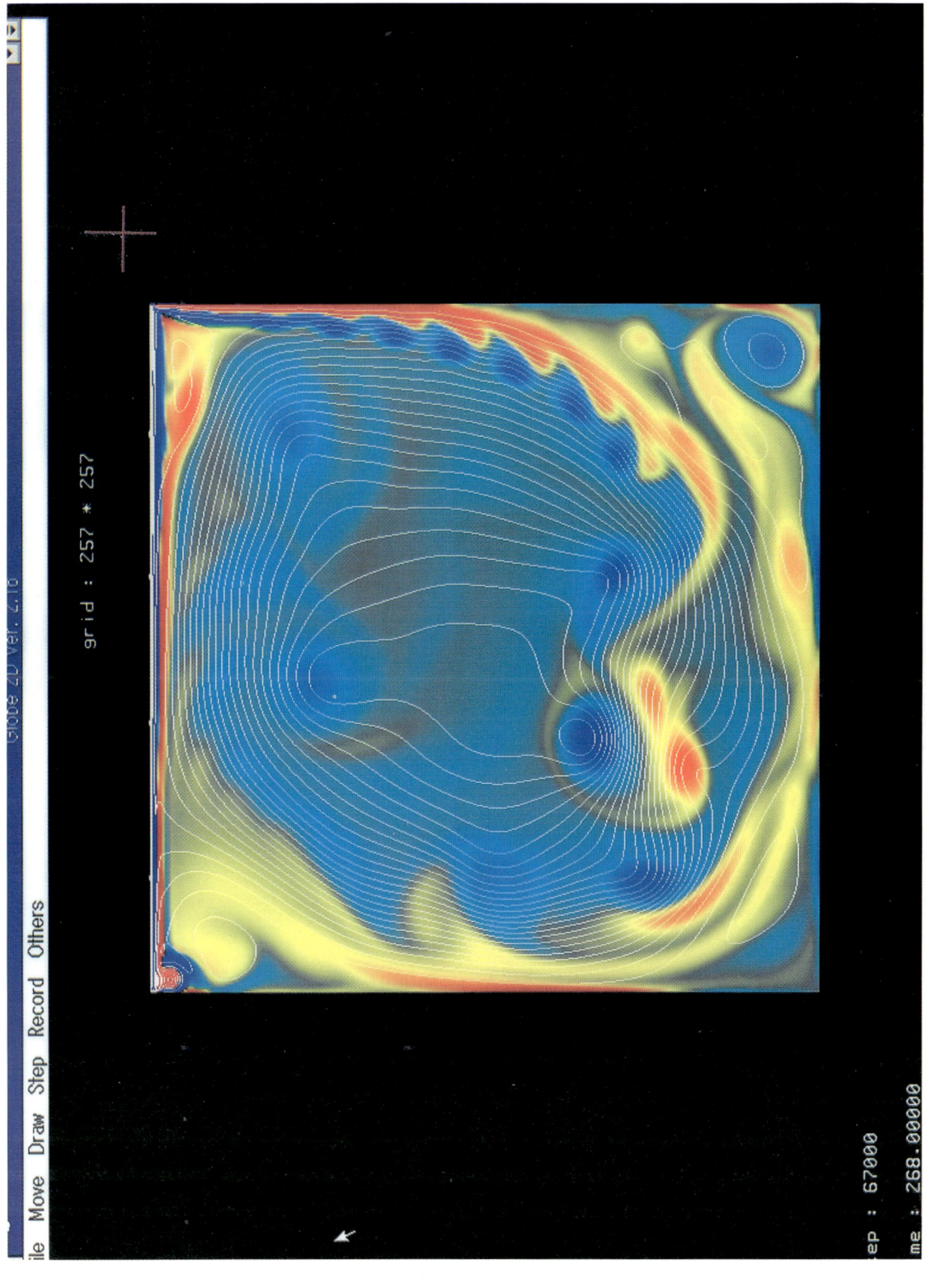

A rare photo of square cavity flow at Reynolds number 100000 after a rather long time continued from Fig.2 (see p.165) : Offered by K.Kuwahara

Lectures on Computational Fluid Dynamics
Mathematical Physics, and Linear Algebra

Karl Gustafson

Contents

Introduction 3

Part I. Recent Developments in Computational Fluid Dynamics
Perspective 7
Chapter 1   Cavity Flow
   1.1. CFD Short Course 8
   1.2. A Computational Benchmark Problem 10
   1.3. Hopf Bifurcation 14
Chapter 2   Hovering Aerodynamics
   2.1. Biological Motivations 17
   2.2. Computational Model 18
   2.3. Vortex Dynamics and Lift 21
Chapter 3   Capturing Correct Solutions
   3.1. Numerical Rotational Release 25
   3.2. Vorticity in the Far Field 30
   3.3. Splitting Errors 36
   Comments and Bibliography 39
   References 47

Part II. Recent Developments in Mathematical Physics
Perspective 54
Chapter 1   Probabilistic and Deterministic Description
   1.1. Theories of Prigogine and Kolmogorov 55
   1.2. Three Converses 61
   1.3. Irreversibility and Second Laws 68
Chapter 2   Scaling Theories
   2.1. Multiresolution Analyses 77
   2.2. Wavelets and Stochastic Processes 83
   2.3. The Time Operator of Wavelets 86
Chapter 3   Chaos in Iterative Maps
   3.1. Attractors and Repellers 89
   3.2. A Gap Theory for Information Dimension 91
   3.3. Onset of Chaos in Neural Learning 102
   Comments and Bibliography 108
   References 112

Part III. Recent Developments in Linear Algebra
Perspective 123
Chapter 1   Operator Trigonometry
   1.1. Semigroup Generators 124
   1.2. Cos A and Sin A 126
   1.3. The Min-Max Theorem 127
Chapter 2   Antieigenvalues

2

2.1. Variational Formulation 133
2.2. The Euler Equation 134
2.3. Kantorovich–Wielandt Inequalities 137
Chapter 3   Computational Linear Algebra
3.1. Optimization Algorithms 141
3.2. Iterative Solvers for $Ax = b$ 146
3.3. Preconditioning Numbers 149
Comments and Bibliography 153
References 156

Introduction

Between October 25 and November 6, 1995, the author presented the following 6 lectures in Japan:

Department of Physics, University of Tokyo, Tokyo, October 26, 1995.

Two New Results in Mathematical Physics: (a) A Gap Theory for Information Dimension, (b) The Time Operator of Wavelets.

National Scientific Research Institute of Aerospace, Tokyo, October 27, 1995.

Hovering Aerodynamics, Cavity flow, and a New Notion of Rotational Release for Unsteady Computational Fluid Dynamics.

Society for Applied Mathematical Analysis of Japan, Tokyo, October 28, 1995.

Antieigenvalues and Their Applications.

Department of Applied Physics and Department of Mathematics, Waseda University, Tokyo, October 30, 1995.

Some Mathematical/Physical Issues in CFD.

National Research Institute of Mathematical Sciences, University of Kyoto, Kyoto, November 2, 1995.

From Probabilistic Descriptions to Deterministic Dynamics: Theories of Kolmogorov and Prigogine.

National Research Institute of Mathematical Sciences, University of Kyoto, Kyoto, November 2, 1995.

Qualitative Features of High Lift Hovering Aerodynamics.

These lecture notes contain the contents of those lectures. In some cases, more will be found here.

These lectures were research oriented and describe current results and issues in certain parts of computational fluid dynamics, mathematical physics, and linear algebra. Accordingly, it was decided to present the material in three separate Parts carrying those names. Each Part begins with a Perspective which focuses the particular questions to be addressed, followed by three (somewhat distinct) chapters, concluding with a Bibliography. In each chapter, roughly, the first section provides the background and motivation for the topic under discussion, the second section presents the current state of the art, the third section gives very newest research results. An attempt to embrace both specialists and beginners will be seen in the style of conceptual explanations mixed with concise results and conjectural discussions.

Generally speaking, Part I addresses mathematical issues of current interest to engineers, Part II addresses mathematical issues of current interest to physicists, Part III addresses mathematical issues of current interest to mathematicians. From the mathematical point of view, Part I is concerned with the Navier–Stokes equation of incompressible fluid dynamics and its computational implementations for the modelling of fluid flow, Part II addresses theories of irreversibility, wavelets, and chaos, and Part III presents a new theory of antieigenvalues for operators and matrices. The

4

two flows studied in Part I are driven cavity flow, for which no analytical solutions are known, and the unsteady motions of dragonflies and hummingbirds, which could one day perhaps usher in a new era in aerodynamics. Part II begins with a statistical ergodic theory of irreversibility developed initially by I. Prigogine, A. Kolmogorov and others. Then it is shown that the currently popular wavelet theory of harmonic analysis may be usefully placed within a context of stochastic processes, and that all wavelets are exactly the eigenstates of a Time operator. Part II concludes with recent results for the fractal dimension of chaotic attractors and repellers and a fundamental occurrence of chaos in neural learning algorithms. Part III presents a theory of antieigenvalues which may be regarded as a significant extension of the usual Rayleigh–Ritz theory of eigenvalues.

It is a great and warm pleasure to thank Professor Takehisa Abe for taking the lead in organizing these lectures and for editing this volume. Thanks also go to Professor Miki Wadachi of the Department of Physics of the University of Tokyo, Professor Kunio Kuwahara of the National Institute of Space and Astronautical Science, Professor Mitsuharu Ohtani and Professor Yoji Aizawa of the Department of Applied Physics and the Department of Physics, respectively, of Waseda University, and Professor Huzihiro Araki and Professor Hisashi Okamoto of the National Research Institute of Mathematical Sciences in Kyoto. I would like to express my gratitude to Kaigai Publishers for undertaking these lectures and their publication, and as well to all others who provided support and friendship during my two weeks in Japan. Finally, last but not least, as usual I am indebted to Elizabeth Stimmel for her conscientious and timely typing of the manuscript.

Karl Gustafson
Boulder
10 January, 1996

Figures and Tables

Fig. I-1. Driven cavity flow, aspect ratio $A = 1$.
      From Gustafson and Leben [1].

Table I-2. Cavity corner stream function intensities.
      From Gustafson and Leben [3].

Fig. I-3. Cavity Hopf bifurcation, $A = 2$, Re $= 10,000$.
      From Gustafson and Halasi [2].

Fig. I-4. Three hovering modes combining plunge and pitch.
      From Gustafson, Leben, McArthur [1].

Fig. I-5. Airfoil domains and coordinate systems.
      From Gustafson and Leben [4].

Fig. I-6. Hovering aerodynamics, experimentally and computationally.
      From Gustafson and Leben [5].

Fig. I-7. Hovering lift C_L and thrust C_T comparisons.
      From Gustafson, Leben, McArthur [1].

Fig. II-1. Loss of quantum probability, a microscopically irreversible event.
      From Gustafson [7].

Fig. II-2. An accretive escaping map.
      From Gustafson [8].

Fig. II-3. Chaos onset in neural learning.
      From Gustafson [9].

Table III-1. Example illustrating combinatorial higher antieigenvalues.
      From Gustafson [12].

Fig. III-2. Example illustrating preconditioned convergence.
      From Sobh and Gustafson [1].

PART I

Recent Developments in Computational Fluid Dynamics

Perspective. In this first part, we will consider mathematical, physical and computational developments and questions concerning two interesting problems in CFD (Computational Fluid Dynamics). The first, cavity flow, is a central benchmark problem often used for testing new code and new computational schemes. In spite of this, many questions remain for this flow. The second problem, hovering aerodynamics, has only recently been studied. It brings out in a nice way many as yet poorly understood issues in unsteady aerodynamics.

CFD as practiced by engineers, physicists, meteorologists, and others, is now a huge enterprise, having begun about 30 years ago. Much of the aerospace industry, the petroleum industry, weather prediction, environmental modelling, now depend on CFD. However, now that computational speed and memory size limitations have been significantly overcome by the computer manufacturers, interest has turned to higher resolution simulations of specific important fluid features. This in turn requires a better understanding of fluids and their mathematical properties.

In Chapter 1 we briefly trace the history of the driven cavity problem, and then present recent results and some current questions. We include theoretical as well as computational developments.

In Chapter 2 we recount the biological motivations for hovering aerodynamics, and recent laboratory experiments, which indicate the very high lift which can be expected from these motions. We then describe the computational simulations which have been performed to date.

In Chapter 3 we discuss certain issues in the relationships of numerical solutions to physical solutions. These include a new concept of numerical rotational release occurring in implicit schemes, the vorticity condition at the far field boundary which is seen to be grid related, and splitting errors as viewed through stencil exponentiations

Chapter 1   Cavity Flow

1.1. CFD Short Course. Before describing the cavity problem, let us establish the general context in which CFD takes place. Imagine fluid flow within any bounded region Ω. This flow is described quite accurately by the Navier–Stokes equations

$$(1.1.1) \qquad u_t - \frac{1}{\mathrm{Re}}\,\Delta u + (u \cdot \nabla)u = -\nabla p + f \quad \text{in } \Omega$$

where u is the velocity vector, $\Delta = \nabla^2$ is the Laplacian operator, Re is the Reynolds number of the flow, p is the pressure, and f represents all externally applied forces. Coupled to (1.1.1) are the initial condition of the fluid, and boundary conditions present during the fluid's time evolution. If the fluid is incompressible (and most of our treatment will assume that it is), one has the constraint

$$(1.1.2) \qquad \nabla \cdot u = 0 \quad \text{in} \quad \Omega$$

There are three 'tricks' which greatly help to understand the history of successful flow modelling developments in CFD. Let us expose them here at the outset. We assume that the fluid is incompressible (div $u = 0$) and the external force solenoidal (div $f = 0$).

Pressure Trick: Take the divergence of the momentum equation (1.1.1). This leads to a potential equation

$$(1.1.3) \qquad -\Delta p = \mathrm{div}\,(u \cdot \nabla)u$$

for the pressure.

Vorticity Trick: Take the curl of the momentum equation (1.1.1). This leads to a heat equation

$$(1.1.4) \qquad \omega_t - \frac{1}{\mathrm{Re}}\,\Delta\omega + (u \cdot \nabla)\omega = 0$$

where $\omega = \nabla \times u$ is the vorticity (i.e., the local twist of the flow).

Stream Function Trick: Introduce a scalar potential ψ for the velocity $u = (u, v)$ by letting $u = \partial\psi/\partial y$, $v = -\partial\psi/\partial x$. This leads to a potential equation

$$(1.1.5) \qquad -\Delta\psi = \omega$$

where in two dimensions $\omega = \partial v/\partial x - \partial u/\partial y$.

All three tricks reduce a vector equation to a scalar equation, a useful simplification. The stream function as described above is two-dimensional. Although recently large scale computers have enabled the simulation of three-dimensional flows, most of our discussion will be two-dimensional, where many of the basic flow features may be observed and studied.

REMARK 1.1.1. The Reynolds number Re in (1.1.1) represents the basic or average speed of the flow and is a nondimensional number, e.g., $Re = UL/\nu$, where U is a driving velocity, L is a typical scale (e.g., the length of a side of the flow region), and ν is the fluid viscosity. Other forms of the Reynolds number must be employed for accelerating flows, see Chapter 2 below. Some tensor linearization has gone into the derivation of (1.1.1) but the important nonlinear term $(u \cdot \nabla)u$ has been retained. For high speed flows sometimes the diffusion term $(1/Re)\Delta u$ is dropped and then one has the so-called Euler Equations. These form a hyperbolic system and have been successfully applied to the design of high speed aircraft. On the other hand, if one drops the nonlinear transport term, one is left with the (linear) Stokes equations. These are often suitable for computing low speed laminar flows.

Early on Lord Rayleigh [1] recognized that when a viscous fluid is in contact with a flow region boundary, certain qualitative features of the fluid exist independent of the magnitude of the viscosity:

> As Sir G. Stokes has shown, the steady motion is the same whatever be the degree of viscosity; and yet it is entirely different from the flow of an inviscid fluid in which no rotation can be generated. Considerations such as this raise doubts as to the interpretation of much that has been written on the subject of the motion of inviscid fluids in the neighbourhood of solid obstacles.

We will take the same point of view: the full Navier–Stokes equations (1.1.1) will be considered. This is especially important for problems exhibiting rich vortex structures.

REMARK 1.1.2. Most CFD solvers are iterative methods dependent upon linear solvers $Ax = b$ in each iteration. For example when the pressure trick (1.1.3) is employed, one alternates back and forth between the pressure equation (1.1.3) and the momentum equation (1.1.1). The condition (1.1.2) is brought in by projection onto the incompressible subspace, or by approximate (so-called artificial viscosity) methods. Solving in pressure p and velocity u variables is called a primitive variables method and works in both two and three dimensions. When solving in the stream function ψ and vorticity ω variables, one alternates between the stream function equation (1.1.5) and the vorticity transport equation (1.1.4). This method is restricted to two space dimensions but generalizes easily to three dimensions, where one of the more effective versions solves in velocity and vorticity variables. Many interesting implementation details enter into these schemes, especially the proper setting of boundary conditions.

REMARK 1.1.3. There is a huge literature on fluid dynamics, and a burgeoning literature already on CFD. We may mention the classic Batchelor [1] for fluid dynamics, Temam [1] for the Navier–Stokes equations, Peyret and Taylor [1] for computational methods. For recent developments emphasizing the vortex dynamics which underlie most fluid flow, see the book Gustafson and Sethian [1], and the references therein.

1.2.   A Computational Benchmark Problem. The driven cavity problem is shown schematically in Fig. I-1. A fluid initially at rest in a two-dimensional cavity region Ω is instantaneously started into motion by the lid of the cavity moving leftward with unit velocity. This causes the development of vortex motions within the cavity. As portrayed in Fig. I-1, a principal vortex 1 gradually consumes most of the region extent. Then secondary vortices 2 occur in the lower corners. These in turn may spin up tertiary vortices 3, and so on.

Most attention in CFD has been directed at the unit cavity. However, very early, Kawaguti [1] observed that a somewhat different motion could be inferred for the depth two cavity. Taneda [1] gave some of the earliest laboratory photos of cavity motion (albeit, for very small Re, essentially Stokes flow) for cavities of several depths. Therefore let $A = \text{depth/width}$ be the aspect ratio of the cavity. Time dependent cavity flows for various aspect ratios A were first computed systematically in Gustafson and Halasi [1]. This paper also contains a reasonably complete summary of some of the main studies of driven cavity flow up to 1985. Generally one may think of the driven cavity problem as a two parameter bifurcation problem (Reynolds number Re and Aspect ratio A) although other parameters (e.g., grid size) will also be seen to naturally enter.

The cavity problem, as distinguished from most theoretical studies which assume a smooth (C^2 or C^1) boundary, is characterized by the presence of its corners. As will be seen below, theory can be modified to provide existence and uniqueness results comparable to the case for smooth boundaries. But it is the corners which give the cavity its interest physically and computationally. Physically, interesting vortex structures develop in the corners. Computationally, the rectangular cavity domain is ideal for testing code written in cartesian coordinates for easy transfer to matrix linear solvers.

Lord Rayleigh [2] became interested in flow in corners:

> The general problem thus represented is one of great difficulty, and all that will be attempted here is the consideration of one or two particular cases. We inquire what solutions are possible such that ψ, as a function of r (the radius vector), is proportional to r^m.

Here Rayleigh has assumed the slow motion linearized steady Stokes equations

$$(1.2.1) \qquad \Delta^2 \psi = 0 \text{ in } \Omega, \quad \psi = \partial\psi/\partial\theta = 0 \text{ on } \partial\Omega$$

for flow in a corner. Here the independent variables are in polar coordinates from the origin placed at the corner, and Δ^2 is the biharmonic operator resulting from insertion of the stream function ψ into (1.1.1). Rayleigh tried a variables-separated solution $\psi(r,\theta) = r^m f(\theta)$ and was not able to fit the boundary conditions unless the corner had angle π or 2π. However by separating variables Rayleigh reduced the partial differential equation (1.2.1) to the ordinary differential equation

$$(1.2.2) \qquad \frac{d^4\psi}{d\theta^4} + [(m-2)^2 + m^2]\frac{d^2\psi}{d\theta^2} + (m-2)^2 m^2 \psi = 0$$

Later Dean and Montagnon [1] employed complex variables and allowed fractional powers r^λ and found further solutions. Then Moffatt [1], [2] increased the trial solution to series $\psi = \sum_{n=1}^{\infty} A_n r_n^\lambda f_n(\theta)$ and looked at solution behavior exactly along the ray $\theta = \frac{1}{2}\alpha$ which bisects a corner angle of α. In an asymptotic approximation he found a velocity

$$(1.2.3) \qquad v \sim r^{-1}(r/r_0)^{\alpha^{-1}\xi_1 + 1} \sin(\alpha^{-1}\eta_1 \ln(r/r_0) + \epsilon)$$

where ξ_1 and η_1 satisfy a trigonometric transcendental system of two equations, r_0 is a length scale, and ϵ is a phase. As $r \to 0$, $v = v(r)$ has an infinite number of zeros r_n, $n = 1, 2, \ldots$. These radii r_n are presumably the distances from the corner out to the centers of an infinite progression of eddies descending into the corner.

Early workers in CFD tested their Navier–Stokes codes by running the driven cavity problem and seeing how many corner eddies could be resolved. See the important studies by Benjamin and Denny [1], Ghia, Ghia, and Shin [1], Schreiber and Keller [1], Gresho, Chan, Lee, and Upson [1], and Glowinski, Keller, and Reinhart [1]. At most three of the predicted cornereddies were found. In Gustafson and Leben [1], [2], [3], by use of multigrid schemes and computationally intensive iteration, first 10, then 21, finally 26 corner eddies were computed, for Stokes flow. See Table I-2. Therefore let us state

PROPOSITION 1.2.1. The linearized driven cavity problem possesses (computationally) at least 25 corner eddies and in principle an ∞ number of them.

Proof. See the references above. An account of the Rayleigh–Moffatt theory, and the assumptions made therein, may be found in Gustafson [2].

REMARK 1.2.2. It is not known how many corner eddies one may expect for the Navier–Stokes equations. Taneda's [1] photographs show at best two. Acrivos and Pan [1] photographed only one, Fuchs and Tillmark [1] at best two. The 25 corner eddies found in Gustafson and Leben [3] occur very close to the radial distances predicted by Moffatt [2]. Moreover, the computed intensity falloffs of the stream function also closely agree with those predicted by Moffatt [2]. However, it is difficult to believe that these intensities, e.g., $\psi_{26} = O(10^{-115})$, can be sustained physically.

Let us now briefly summarize what is known theoretically for steady solutions for the driven cavity problem. If we consider the steady viscous incompressible Navier–Stokes equations for a general bounded smooth domain Ω

$$-\frac{1}{\text{Re}} \Delta u + (u \cdot \nabla)u = -\nabla p + f(x) \quad \text{in} \quad \Omega \ ,$$

(1.2.4)
$$\nabla \cdot u = 0 \quad \text{in} \quad \Omega \ ,$$

$$u = g(x) \quad \text{on} \quad \partial\Omega \ ,$$

where, for simplicity, we restrict to the Dirichlet Boundary data $u = g$, then for all Reynolds numbers $0 < \text{Re} < \infty$, there exists at least one solution (u, p). Sufficient conditions for this are

(1.2.5)
$$f \in W^{1,2}(\Omega), \quad g \in W^{(1/2),2}(\partial\Omega), \quad \int_{\partial\Omega} g \cdot n = 0,$$

where $W^{-1,2}$ is the distribution space dual to the Sobolev space $W^{1,2}$ on Ω, and $W^{(1/2),2}$ are the generalized boundary values corresponding to $W^{1,2}$.

The pressure p is only determined up to its gradient. Note, however, that for a given steady solution u, all corresponding pressure solutions p can differ only by a constant. There may be more than one steady solution u. However, for bounded domains Ω, there exist constants C_1 and C_2 depending only on Ω such that, if

$$(1.2.6) \qquad \|f\|_{-1,2} \leqq \frac{1}{(\text{Re})^2}\, C_1 \quad \text{and} \quad \|g\|_{(1/2),2} \leqq \frac{1}{\text{Re}}\, C_2,$$

then there is exactly one solution u in $W^{1,2}(\Omega)$. In other words, for small enough forcing data, there is only one resultant velocity profile in Ω. As Reynolds number increases, the allowable forcing data must be smaller for uniqueness of solutions. For no forcing data at all, the only solution is no motion, $u = 0$, and constant pressure, $p = c$.

These results are well known, e.g., see Temam [1], Ladyzhenskaya [1]. However, their proofs usually require a smooth boundary, with no corners. The theory for domains with corners came later and is treated extensively in Grisvard [1]. The data assumptions and resulting solutions spaces change somewhat, needing modification near the corners. Serre [1] treated specifically the driven cavity domain (and that of the cylindrical Taylor problem) within a context of open bounded connected domains Ω with $\partial\Omega$ composed of a finite number of piecewise smooth (C^2) components Γ^k, $1 \leqq k \leqq N$. When

$$(1.2.7) \qquad f \in W^{-1,p}(\Omega), \quad g \in W^{1-(1/p),p}(\Omega), \quad \int_{\Gamma_k} g \cdot n = 0,$$

$0 \leqq k \leqq N$, $n/2 < p < 2$, there exists a steady solution $u \in W^{1,p}(\Omega)$, $p \in L^p(\Omega)$. Also one can conclude uniqueness of u for small data

$$(1.2.8) \qquad \|f\|_{-1,p} \leqq \frac{1}{(\text{Re})^2}\, C_1 \quad \text{and} \quad \|g\|_{1-(1/p),p} \leqq \frac{1}{\text{Re}}\, C_2$$

as before. Thus, one can treat domains with a finite number of corners, and even with slightly less regular data, if one accepts a slightly less regular solution.

In particular, then, for the driven cavity problem, in which $f = 0$, and $g = 0$ on three sides and $g = (-1, 0)$ or $(1, 0)$ on the top, depending on whether the lid is being driven leftward or rightward, we are assured of the existence of a steady solution to the continuous problem, for all Reynolds numbers $0 < \text{Re} < \infty$. From the above theory, we can assert the uniqueness of the steady solution only for Re in some finite range. Intuitively it would seem reasonable that there might be a unique steady solution for all Re. But, on the other hand, by analogy with other fluids problems in which multiple steady solutions can exist, albeit some more stable than others, it is conceivable that nonuniqueness could occur. Intuitively one can imagine this better in the three-dimensional driven cavity problem, where Taylor–Görtler-like vortices can arise orthogonal to the $2d$ driven cavity sections.

1.3. Hopf Bifurcation. For some time it was believed that for the constant velocity driven lid, one could compute only steady solutions. Recall that u represents the velocity in (1.1.1); thus a steady (fluid) solution is one which is moving, but without acceleration. The early studies discretized only the (easier) steady problem. Then when larger faster computers became available and time dependent solutions could be computed, for the Reynolds numbers for which solutions could be computed, all computed time dependent solutions tended to steady solutions. See Ghia, Ghia and Shin [1], Schreiber and Keller [1], Gresho, Chan, Lee, and Upson [1], and Glowinski, Keller, and Reinhart [1]. However in Gustafson and Halasi [2] by going to the depth 2 cavity, it was demonstrated for the first time that the driven cavity problem possesses a (computational) Hopf bifurcation. This flow at its final stages is shown in Fig. I-3. Later studies have confirmed this result. Hence

PROPOSITION 1.3.1. The driven cavity problem possesses (computationally) a Hopf bifurcation for aspect ratio $A = 2$.

Proof. See Gustafson and Halasi [2]. Many later studies have confirmed this result, see in particular Goodrich, Gustafson, and Halasi [1], Fortin, Jardak, Gervais, and Pierre [1], Shen [1], Liu, Fu, and Ma [1], among others. It should be mentioned that the period of the final oscillation depends on grid size: in Fig. I-3 the period is about 4.5 seconds. On a slightly finer grid, Goodrich, Gustafson, Halasi, [1], the period is about 3.5 seconds. It is an interesting question as to what the limiting period should be as grid size tends to zero.

REMARK 1.3.2. The analytically inured may find it incomprehensible that a computational result be called a proposition. However, in computational practice, where analytic solutions are not available, and where numerical analysis rigorous proofs cannot be or have not yet been available, the proof becomes: show the result to be scheme independent, i.e., obtainable by different numerical schemes, show it to be parameter independent, i.e., to hold qualitatively regardless of (acceptably fine, in this case) grid size employed; and so on. The result of Proposition 1.3.1 generated considerable interest in the CFD community. On the other hand, notice that we have been careful to state that the result is computational: to our knowledge, no analytic proof of the Hopf bifurcation for the cavity has been obtained. Compare Marsden and McCracken [1].

REMARK 1.3.3. It is almost trivial to produce computational periodic motions in the cavity if one is allowed to periodically drive the lid. An early paper demonstrating this is Duck [1]. Recall that a Hopf bifurcation is one which reflects the inherent structure of the equations and which occurs under steady forcing.

Now let us summarize what is known theoretically for unsteady solutions. For the unsteady Navier–Stokes equations for general bounded smooth domains Ω

$$
\begin{aligned}
u_t - \frac{1}{\mathrm{Re}}\,\Delta u + (u\cdot\nabla)u &= -\nabla p + f \quad &&\text{in}\quad \Omega\\
\operatorname{div} u &= 0 \quad &&\text{in}\quad \Omega\\
u &= g \quad &&\text{on}\quad \partial\Omega\\
u(x,0) &= u_0,
\end{aligned}
$$

(1.3.1)

where the initial condition u_0 is required to be divergence-free, then, for all $0 < \mathrm{Re} < \infty$, there exists at least one (weak) solution on any given finite time interval $0, T)$. Of course, there are certain function space conditions needed: for example,

$$
f \in W^{1,\infty}((0,T); W^{-1,2}(\Omega)), \quad g \in W^{1,\infty}((0,T); W^{(1/2),2}(\Omega))
$$

(1.3.2)
$$
u_0 \in L^2(\Omega), \quad \operatorname{div} u_0 = 0, \quad u_0 \cdot n = g(x,0)\cdot n \text{ on } \partial\Omega
$$
$$
\int_{\partial\Omega} g \cdot n = 0 \quad \text{for all}\quad 0 \leqq t \leqq T.
$$

Then a weak velocity solution $u \in L^\infty((0,T); L^2(\Omega)) \cap L^2((0,T); W^{1,2}(\Omega))$ exists, along with a pressure $p \in L^2((0,T); L^2(\Omega))$. These results may be found in the references that we have already mentioned. See Temam [2] for an uptodate account.

However, theory gives thus far very little detail about the flow's intermediate and long term behavior, and to better understand these dynamical flow systems, large scale scientific simulations in the last twenty years have been invaluable. For example, these studies indicate that the Hopf bifurcation for the depth 2 cavity first occurs at some Re between 2000 and 5000. For the depth 1 cavity, the critical Re is apparently between 8000 and 10,000.

REMARK 1.3.4. See Gustafson [3] for a rather detailed description of the full depth 2 driven cavity dynamics and the emergence and history of all vortex events occurring therein. See Gustafson [4] for a recent account of progress on the driven cavity problem. There it is conjectured that the Hopf bifurcation criticality curve, plotted with aspect ratio A horizontal and Reynolds number Re vertical, increases monotonically as one comes to the left from $A = \infty$, rising to a cusp at some A slightly less than one, thereafter decreasing monotonically as A goes toward zero. Recently Fortin, Jardak, Gervais, and Pierre [1] have investigated techniques for finding such criticalities. See also the methods discussed in the account by Winters [1].

REMARK 1.3.5. All of the above discussion has been within the two-dimensional cavity context. Three-dimensional cavity flow experiments (indeed, in the laboratory, all are three-dimensional) have revealed the above two-dimensional cavity behaviors, seen as sections. However, there are interesting three-dimensional effects not accounted for in the above discussion. Soon the three-dimensional driven cavity should be studied in more detail as the next-generation CFD benchmark problem.

Chapter 2   Hovering Aerodynamics

2.1. Biological Motivations. Birds, when in the gliding, straightforward, or soaring modes, succeed to fly by the principles of steady aerodynamics: a uniform motion into a uniform freestream flow. These are also the aerodynamics used in conventional airliner flight: lift created as a result of conservation of circulation of smooth flow of air over wing surfaces. In the design and practice of such flight, every effort is made to avoid or minimize flow separation: hence the creation in our vocabulary of the word 'streamlining.'

Birds, and many insects, when in their hovering modes of landing, fighting, or taking food, exhibit a far more interesting aerodynamics. This hovering, by use of very rapid wing oscillatory motion, has been found to create and utilize special patterns of separation and vortex development to generate the high lift necessary for sustaining such motion. Such aerodynamics is most dramatically seen in the hovering and direction change motions of hummingbirds and dragonflies. In contrast to conventional steady aerodynamics, a very sophisticated unsteady biological dynamical system has evolved for over 100 million years to enable such flight. Only in the last ten years has this hovering dynamics been simulated and computed mathematically. In this chapter we shall discuss these recent developments.

Interest in hovering flight was rekindled by the biological studies of Somps and Luttges [1]. Dragonflies possess four wings and a number of related questions were discussed there. However, hummingbirds possess only two wings and exhibit a similar aerodynamics. Physical experiments with a single mechanical wing by Freymuth [1], [2] demonstrated that a single airfoil executing rapid plunging and pitching can generate a high thrust in the form of a reverse Karman vortex street. In a preliminary joint study, Freymuth, Gustafson, and Leben [1], we developed a computational model and found excellent agreement between the vortical signatures of the experimental visualizations and those of our Navier–Stokes simulations. As reported in Gustafson and Leben [5], Gustafson, Leben and McArthur [1], we successfully simulated the hovering modes in terms of vortex patterns and comparisons of laboratory measured coefficients of thrust, C_T, and computationally derived lift coefficients, C_L. The correlations between the laboratory data and our computational results are in close enough agreement that we believe our computational model provides a significant start for a comprehensive mathematical model of the complicated unsteady motion observed in such animal flights.

Three generic modes are diagrammed in Fig. I-4. Mode 1 or 'water treading mode' is described by an average angle of attack coincident with the horizontal plunging motion, $\alpha_0 = 0$, with a phase angle $\phi = -\pi/2$ $(-90°)$ producing a hover-jet directed downward. In this mode leading and trailing edges switch their role during one cycle of oscillation. Mode 2 or the "degenerate figure eight mode" is characterized by an average angle of attack at a right angle with the horizontal plunging motion, $\alpha = \pi/2(90°)$, with a phase angle $\phi = \pi/2(90°)$ producing a hover-jet directed downward. In this mode leading and trailing edges do not switch their roles during one cycle. This mode resembles the wing motion of hummingbirds and certain flying insects. Mode 3 is an oblique mode described by an average angle of attack oblique to the horizontal plunging motion. In this mode the jet is directed obliquely to the plunge (stroke) plane. Mode 3 combines facets of both Mode 1 and Mode 2 and resembles dragonfly flight.

REMARK 2.1.1. For a general treatment of vortex phenomena including bird and insect flight, see Lugt [1]. An excellent recent book which has just come to our attention is Azuma [1]. There were numerous earlier studies before Somps and Luttges [1], and Azuma [1] comments on these. Roughly speaking, however, most studies before those at the University of Colorado during 1985–1995 avoided dealing directly with the full unsteady aerodynamic models. For example Azuma [1, p. 120] points out in the analysis of beating flight that some conclusions may be obtained by power or energy balances, but that numerical simulation from the partial differential equations would be needed to avoid an assumption of constant induced velocity. In his analysis of dragonfly kinematics Azuma [1, p. 141] uses a local circulation method (LCM), Azuma and Watanabe [1], based upon classical blade element analysis modified to include nonconstant induced velocity distributions. Lift forces lower than those of Somps and Luttges [1] are found. However the LCM method although incorporating unsteady effects remains based upon the potential equation approach to aerodynamics and thus apparently misses the high lift vortex shedding events which we were able to compute, to be discussed in the next two sections.

2.2. Computational Model.

Computation of hovering flight modes presents significant challenges to the computational fluid dynamicist. First, proper formulation of the rotating and translating reference frame, including suitable application of the outflow boundary conditions, is necessary to properly resolve the patterns of the self-induced vortical flow. Second, accurate body force computations are necessary to estimate the aerodynamic efficiency of such flight. Third, care must be given to proper scaling for correct validation of the numerical simulations with the physical laboratory photography. Our method, following our earlier work on flow about accelerating airfoils, Gustafson and Leben [4], conformally maps the exterior of an elliptical wing cross-section into a circular interior domain, which is then numerically mapped to a depth-2 cavity with a uniform grid to facilitate efficient linear solvers. See Fig. I-5. By this approach we obviate the need to specify outflow boundary conditions at some far-field computational boundary in the physical domain. Additionally, we concentrate the finite difference grid near the wing to enable high resolution of the vortex shedding which has been found to be essential to hovering aerodynamics. Thus we sacrifice far field detail resolution in order to obtain accurate simulation of the near field dynamics, from which we also compute the lift coefficient C_L.

The physical simulation of the hovering modes by Freymuth [1], [2] was achieved by the rapid airfoil movement in still air. In our numerical simulations, the reference frame is fixed with the airfoil. To allow simulation of the hovering mode, the freestream flow relative to the airfoil is rapidly varied about the airfoil. We nondimensionalize the angular rotation rate of the airfoil as:

$$(2.2.1) \qquad \Omega = \frac{a\dot{\alpha}}{U}$$

where

$a \equiv$ characteristic length (approximately one-half the chord length c)

$\dot{\alpha} \equiv$ angular rotation rate of the airfoil

$U \equiv$ translational velocity at midcycle.

The stream function due to the flow at infinity is modified to include the effect of the angular rotation and translation:

$$(2.2.2) \qquad \psi_\infty = \psi_{\text{freestream}} + \psi_{\text{rotation}}$$

Thus,

$$(2.2.3) \qquad \psi_\infty = U_{\text{translational}}(\cos(\alpha)Y - \sin(\alpha)X) + \frac{\Omega}{2}\left(X^2 + Y^2\right)$$

Corresponding to the orthogonal uniform grid in the rectangular computational domain, a near-orthogonal grid is algebraically generated in the auxiliary circular interior domain. This is conformally mapped back to the exterior physical domain to determine a near-orthogonal ξ, η grid on the infinite exterior physical domain. The equations of motion are formulated in terms of the stream function ψ and ω, the vorticity component normal to the ξ, η plane. Since the vorticity field observed in the reference frame fixed with the hovering airfoil differs only by a constant from one at rest, we introduce the disturbance stream function:

$$(2.2.4) \qquad \psi^* = \psi - \psi_\infty.$$

Then, the governing equations to be solved for comparison to the flow visualization experiments are

$$(2.2.5) \qquad \frac{\partial \omega}{\partial t} + \frac{1}{h_1 h_2} \left(\frac{\partial \tilde{u}\omega}{\partial \xi} + \frac{\partial \tilde{v}\omega}{\partial \eta} \right) = \frac{L}{R} \, \nabla^2 \omega$$

and

$$(2.2.6) \qquad -\nabla^2 \psi^* = \omega$$

where

$$(2.2.7) \qquad \tilde{u} = \frac{\partial \psi^*}{\partial \eta} + \tilde{u}_\infty = h_2 u, \qquad \tilde{v} = -\frac{\partial \psi^*}{\partial \xi} + \tilde{v}_\infty = h_1 v,$$

and

$$(2.2.8) \qquad \begin{aligned} \tilde{u}_\infty &= h_2 u_\infty = \frac{\partial \psi_\infty}{\partial \eta} \\ \tilde{v}_\infty &= h_1 v_\infty = -\frac{\partial \psi_\infty}{\partial \xi}. \end{aligned}$$

and where

$$(2.2.9) \qquad \nabla^2 = \frac{1}{h_1 h_2} \left[\frac{\partial}{\partial \xi} \left(\frac{h_2}{h_1} \frac{\partial}{\partial \xi} \right) + \frac{\partial}{\partial \eta} \left(\frac{h_1}{h_2} \frac{\partial}{\partial \eta} \right) \right].$$

Here L is the dimensionless chord length and R is the Reynolds number which must be chosen to match that used in the physical laboratory. Boundary conditions are no slip velocity on the airfoil surface, irrotational flow at infinity, a first order differencing of the Laplacian of the streamfunction for the vorticity at the airfoil surface, and an impermeability condition for the streamfunction at the airfoil boundary.

Now consider a thin airfoil with chord length c exposed to still air and executing a translating (plunging) motion h in horizontal direction:

$$(2.2.10) \qquad h = h_a \sin(2\pi f t)$$

where h_a is the amplitude of translation, f is the frequency of oscillation and t is time. Consider the airfoil to simultaneously execute a pitching motion around the half-chord axis:

$$(2.2.11) \qquad \alpha = \alpha_0 + \alpha_a \sin(2\pi f t + \phi)$$

where α is the angle of attack with respect to the horizontal line, α_0 is the average angle of attack, α_a is the pitch amplitude and ϕ is the phase difference between pitching and plunging. The dimensionless parameters of the system are: α_0, α_a, ϕ, the dimensionless plunge amplitude h_a/c and a Reynolds number $R_f = 2\pi f h_a c/\nu$ based on maximum plunge speed and on c, where ν is the kinematic viscosity which for air is taken as 0.15 cm/sec^2. The airfoil used in the physical experiments with which we compared was of cross section 25.4 mm $\times$ 1.6 mm, with slightly rounded edges, and long span to provide two-dimensional midwing dynamics for photographic visualization. Flow history is recorded by placing on the upper surface of the airfoil a liquid (e.g., titanium tetrachloride TiCl$_4$) which reacts with air to give off a dense white smoke, thereby tagging and making visible (in white on black) the vortex patterns of the flow. This technique gives excellent visualization in the middle flow field, but is overly bright in the very near field, and probably departs somewhat from the actual fluid vorticity in the far field.

For our hovering mode computational comparisons, we used an elliptical airfoil of approximately 10% thickness ratio to represent the near-rectangular laboratory airfoil. The slight differences, e.g., in roundedness of airfoil ends, between the laboratory and computational airfoils were found to be essentially acceptable for the comparisons. The hovering mode computations usually used a 33 $\times$ 65 or a 65 $\times$ 65 mesh. Computational flow visualization of the vorticity field was performed by means of a multigrid interpolation scheme which produced byte files for display.

The covariant Poisson equation (2.2.6), (2.2.9) was solved by multigrid, and the vorticity transport equation (2.2.5), (2.2.9) was solved by an ADI time marching scheme. The coefficient matrices in the (perpendicular to airfoil) ξ direction are tridiagonal, in the η direction (which is streamwise parallel to the airfoil) after some experimentation we went to periodic line relaxation in order to resolve the far field.

REMARK 2.2.1. There are a number of essential modelling and code implementation details which are not discussed here. See Gustafson and Leben [2,3,4,5] for more information.

2.3. Vortex Dynamics and Lift. The excellent correspondence found between the vortex

22

patterns of the mathematical computer simulation (on the right) and the physical laboratory photographs (on the left) is shown in Fig. I-6. There the computer simulation display frame is slightly larger than that of the physical laboratory photographs, and the computer simulation is slightly behind the physical dynamics in time. Nonetheless remarkable agreement of near-field vortex patterns as calculated from the Navier–Stokes equations according to the model in the preceding section, with those from physical laboratory photographs, was found in all cases compared, for all three modes 1, 2, and 3 of Fig. I-4. Essentially all available physical laboratory sequences have been matched. See Freymuth, Gustafson and Leben [1], Gustafson and Leben [5], Gustafson, Leben and McArthur [1], for further vortex dynamics simulations.

Encouraged by these results, we then calculated the lift coefficients C_L for comparison to the downstream thrust coefficients C_T calculated by Freymuth [2]. In the experimental studies of Freymuth, large time averaged thrust coefficients C_T (e.g., up to 6) were found in well tuned hovering modes. These coefficients are measured experimentally from the formula

$$(2.3.1) \qquad C_T = \int_{-\infty}^{\infty} \bar{V}^2 dx / (\pi f h_a)^2 c$$

where $\bar{V}^2$ is the mean square velocity at a sufficient distance downstream where the vortical jet signature has acquired ambient pressure. These mean velocities were obtained by averaging a cross section of pitot tube measurements, usually four chord lengths downstream. Should $C_L \approx C_T$ approximately, the aerodynamic implications are interesting. In steady commercial flight, lifts seldom exceed $C_L \equiv 2$.

In order to verify the high coefficients of thrust claimed for the hovering airfoil model, we executed a set of numerical simulations at the identical hover-jet parameter values used for the experimental calculation performed at the maximum frequencies allowed by the experimental apparatus; $R_f = 1700$ and $f = 1.8$. The calculation of an average thrust coefficient from the time averaged far field momentum excess is equivalent to an average of the instantaneous lift coefficient computed by an integration of the airfoil surface pressure and friction forces perpendicular to the stroke plane. The pressure at the airfoil surface is given by:

$$(2.3.2) \qquad p(\eta) = \frac{L}{R} \int_0^{\eta} \frac{h_2}{h_1} \frac{\partial \omega}{\partial \xi} \, d\bar{\eta}.$$

The coefficient of lift for the airfoil is the sum of pressure and friction components given, respectively, by:

(2.3.3)
$$C_{L_p} = -\frac{1}{L^2\pi f h_a/c^2}\left[\cos\alpha\int_0^1\frac{\partial x}{\partial\eta}\,p\,d\eta + \sin\alpha\int_0^1\frac{\partial y}{\partial\eta}\,p\,d\eta\right].$$

and

(2.3.4)
$$C_{L_f} = -\frac{1}{RL\pi f h_a/c^2}\left[\cos\alpha\int_0^1\frac{\partial y}{\partial\eta}\,\omega\,d\eta - \sin\alpha\int_0^1\frac{\partial x}{\partial\eta}\,\omega\,d\eta\right].$$

The nondimensionalization $\rho\,\overline{V(t)^2}_{\text{translation}}c/2$ has been used for the lift and thrust coefficients, where ρ is the air density, and $\overline{V(t)^2}_{\text{translation}} = (2\pi f h_a)^2/2$.

Let us comment further on some considerations of nondimensionalization and proper scaling for correct comparison of physical experiment with computer simulation. The laboratory measurements of Freymuth were generated using the chord length for nondimensionalization. In our simulations the focus length of the ellipse was used for nondimensionalization, hence the factors of L appearing in the equations for lift coefficients. Appropriate modifications were made to the other affected quantities. After an initial startup of four plunging cycles, averages of the lift coefficients were taken over cycles five through eight for each hovering mode simulation parameter setting.

For mode 1 hovering, our lift coefficient computations are in remarkable agreement with Freymuth's thrust calculations. Plunge amplitudes at which peak thrust and lift coefficients occur agree for all of the angles of attack investigated. The results of this rather extensive parameter study are presented in Fig. I-7.

Results for mode 2 hovering do not compare as favorably. While Freymuth found thrust coefficients to be larger for well-tuned mode 2 hovering as compared to well-tuned mode 1 hovering, our numerical results show nearly equal peak lift coefficients for the two hovering modes. The qualitative trends in the mode 2 lift and thrust coefficients are in agreement, however, the experimental calculations predict consistently larger aerodynamic efficiency for this mode which is not seen in the numerical results. Possible explanations for these differences are currently under study. There were not sufficient mode 3 physical laboratory measurements with which to compare. Mode 3 presents an interesting problem in the laboratory of trying to determine appropriate positions for pitot tube velocity measurements in an oblique jet of counterrotating shed vortex pairs.

See Gustafson, Leben and McArthur [1] for more details on these lift computations.

We may summarize as follows.

PROPOSITION 2.3.1. *Mathematically simulated hovering dynamics agrees qualitatively (vortex dynamics) with the physical dynamics. Quantitative agreement between the mathematically computed lift coefficient C_L representing surface pressures and the physical downstream wake thrust coefficient C_T has also been obtained.*

REMARK 2.3.2. Further studies of qualitative features of hovering dynamics, e.g., phase portraits of the instantaneous lift curves $C_L(t)$ and their power spectra, the possibilities of period doubling to aperiodic behavior to possible chaos as the Reynolds number parameter increases, future useful relationships of hovering dynamics as related to attractor and inertial manifold theory, may be found in Gustafson, Leben, McArthur and Mundt [1]. Further discussion of the importance of the related bioaerodynamical subsystems (thermodynamical, structuraldynamical, controldynamical) to a better understanding of the feasibilities of aircraft design employing the high lift features of hovering aerodynamics may be found in Gustafson [5].

REMARK 2.3.3. Although much of the basic hovering aerodynamics can be clearly understood by the two-dimensional model developed to date, a further step would be to study the full three-dimensional model. Also, we have not emphasized the drag coefficient C_D, which is easily computed similar to C_L above, and its comparisons to C_L. Interesting comparisons to helicopter design models could be carried out.

Chapter 3   Capturing Correct Solutions

3.1. Numerical Rotational Release. To tolerate larger Reynolds numbers in computing solutions to the Navier–Stokes equations, a well known and often used remedy has been to replace explicit schemes by implicit ones. For example, such implicit schemes result by the use of spatial upwind differencing, which eliminates unwanted numerical oscillations, and by the use of temporal backward differencing, which prevents unstable solution growth. Such implicit iterative schemes have the advantage of converging. However, because they suppress oscillation and/or rapid growth, they will generally converge to the steady, or steadier, solution. Thus as CFD progresses up the bifurcation ladder beyond computing just steady solutions, to computing periodic or aperiodic solutions, care must be taken to ascertain that a numerical scheme will capture the 'correct' solution.

The following is taken from Gustafson [6], which grew out of a conversation with G. Kobelkov in Moscow in 1992. G. Kobelkov [1], see also earlier papers referenced there, introduced a scheme for the numerical computation of the full viscous incompressible Navier–Stokes equations in which there enters a term created by an algebraic substitution in order to make the scheme fully implicit. This scheme is advantageous in that it computes solutions to the Navier–Stokes equations for large (e.g., Re = 100,000) Reynolds numbers. On the other hand, at Re = 10,000 for the depth 2 cavity it finds steady solutions, whereas in Gustafson and Halasi [2] periodic solutions were obtained, as was discussed in Section 1.3.

One may physically interpret the terms in Kobelkov's [1] scheme and from this we wish to introduce here a new notion of *numerical rotational release*, Gustafson [6]. Possibly this new concept of numerical rotational release should be viewed, as is numerical viscosity, as a fundamental numerical phenomenon useful throughout CFD.

In its simplest setting, that of the stationary Stokes problem

(3.1.1)
$$\begin{cases} -\Delta u + \nabla p = f & \text{in } \Omega \\ \operatorname{div} u = 0 & \text{in } \Omega \\ u = 0 & \text{on } \partial\Omega, \end{cases}$$

Kobelkov's iterative scheme [1] may be written as

(3.1.2)
$$B(u^{n+1} - u^n)/\tau - \Delta u^{n+1} + \nabla p^{n+1} = f,$$
$$\beta\tau(p^{n+1} - p^n)/\tau + \operatorname{div} u^{n+1} = 0.$$

Letting $u_t^n = (u^{n+1} - u^n)/\tau$ and $C = (B - \tau\Delta - \tau\beta^{-1}\nabla\mathrm{div})$, after algebraic manipulation the scheme may be written as

$$Cu_t^n = \Delta u^n + \beta^{-1}\nabla\mathrm{div}\,u^n - \nabla p^n + f,$$

(3.1.3)

$$p^{n+1} = p^n - \beta^{-1}\mathrm{div}\,u^{n+1}.$$

The term of interest is $\beta^{-1}\nabla\mathrm{div}\,u^n$. Without this term, the scheme may be thought of as similar to the Uzawa–Arrow–Hurwitz schemes (cf. Temam [1]).

To fix ideas, let us first ignore the fact that the term of interest also occurs in the operator C, indeed, C may be taken to be the identity I in a number of instances. Thus we see that the iterative scheme (3.1.3) has converted the given stationary Stokes problem into a nonstationary problem. As is evident, the basic idea is to initially permit some compressibility, which is then to be driven to zero as the pressure iterations converge. The preconditioner B should be kept symmetric positive definite, from which convergence of the scheme ensues. For example, with $C = -\Delta$, B becomes $B = -(1 - \tau)\Delta + \tau\beta^{-1}\nabla\mathrm{div}$, which will be positive provided that $\tau(1 + \beta^{-1}) < 1$.

It is useful now to remember the Helmholtz–Weyl vector field decomposition (cf. Temam [1] or Gustafson [1]) $u = u_1 + u_2 + u_3$ according to

(3.1.4) $$(L^2(\Omega))^n = H^1 \oplus H^2 \oplus H^3$$

where $\mathrm{div}\,u_1 = \mathrm{curl}\,u_1 = 0$; $\mathrm{div}\,u_2 = \mathrm{curl}\,u_2 = \mathrm{curl}\,u$; $\mathrm{div}\,u_3 = \mathrm{div}\,u$, $\mathrm{curl}\,u_3 = 0$. Classically, this is just the view of an arbitrary smooth vector field decomposed into its solenoidal and irrotational parts, the former determined uniquely by its vortices, namely the points in the region at which its curl is nonzero, the latter determined uniquely by its sources, namely the points in the region at which its divergence is nonzero. However, within a specific finite element or finite difference scheme, such a decomposition can take on special meanings.

Turning first to the pressure iteration in (3.1.3), we may introduce the suggestive name *divergence pressure* for the irrotational component $\beta^{-1}\mathrm{div}\,u_3^{n+1}$. Of course, divergence pressure is, numerically speaking, a kind of artificial compressibility and in that sense is not new. However, within the schemes of Kobelkov [1], divergence pressure will be seen to carry three implications. First, trivially, it is indeed a pressure increment, caused by a computationally designed relaxation of strict incompressibility. Second, at points in the computational domain of the fluid where it is significantly nonvanishing, the divergence pressure term connotes local pressure expansions of contractions of the computational fluid. For example, when $\operatorname{div} u^{n+1} < 0$, we see in (3.1.3) that pressure is to increase. Thus, numerically induced artificial velocity field sinks are to be compensated by raising the pressures there, whereas numerical velocity field sources will be compensated by reducing pressures there. Geometrically, divergence pressures thus have the effect of 'flattening' the flow. The third implication of divergence pressure is a physical interpretation of Kobelkov's fully implicit term, $\beta^{-1}\nabla\operatorname{div} u$. This term is the spatial gradient of the divergence pressures. The physical interpretation of this term is what we shall call (numerical) *rotational release*. In iterative schemes such as those of Kobelkov [1], although rotational release is the implicit mechanism which enables convergence at higher Reynolds numbers, the price that is paid is a tendency toward convergence to steady solutions, when in fact the physical flow would prefer a qualitative change toward periodic solutions.

Let us note that for very small pseudo time step τ in the schemes (3.1.2), (3.1.3), for the mentioned preconditioner B to be positive, the parameter β must exceed a (small) positive lower bound. Thus, even for small pseudo time step, there may be significant numerical divergence pressures, and more important, significant numerical rotational release. For the computational efficiencies of larger pseudo time step τ, β must exceed a (larger) positive lower bound, so β^{-1} can be smaller, but presumably the divergence pressures and rotational release effects would themselves be initially larger in that case. Whatever be the case, the strength of iterative implicit schemes such as (3.1.2), (3.1.3) are that eventually these pressure divergence anomalies will be balanced out and eventually eliminated altogether, leading to convergence even at higher Reynolds numbers.

Thus, the (false) divergence pressures of an implicit scheme can create a selection of steady rather than unsteady solution. While in process, an iterative implicit scheme effectively Re-continues these steady solutions and thereby may miss a key exchange of stabilities phenomenon, in the case of the depth 2 cavity flow of Gustafson and Halasi [2], the generation of the sidewall tertiary eddies, which in nature enables the transition to qualitatively more complex flows. Inasmuch as steady flows do exist (theory shows this, see Section 1.2) for the driven cavity at all Reynolds numbers, Kobelkov's fully implicit scheme [1] may, or may not, be finding exactly those steady solutions. As shown in Temam [1], generally a method of artificial compressibility, as implemented by implicit schemes of fractional steps, may be expected to converge to the true solution of the Navier–Stokes equation, although the considerations of Temam [1] consider only domain forcing, and not the boundary forcing that we have here.

Rotational release is the spatial rate of failure to capture small eddies. Totally rotation-free itself, it represents the gradient of small implicit numerical divergences which permit the escape of fluid from small recirculatory regions in which it is essential that constrained eddy motion form. Mathematically it represents an allowed numerical shift from the subspace H^2 to the subspaces H^1 and H^3 of (3.1.4). As a penalty or relaxation or implicit term inserted to make a numerical scheme more robust, it has the side effect which may permit a flow to leave a region in which a rotation should take place. Pairs of such small counterrotating regions are essential for periodic fluid motions, as they permit the accelerative local surges inherent in such unsteady motions.

To better understand this new notion of rotational release, let us consider the Navier–Stokes equations

$$(3.1.5) \qquad u_t - \frac{1}{\mathrm{Re}} \Delta u + (u \cdot \nabla)u = -\nabla p.$$

For simplicity, we restrict to two dimensions. Employing the pressure trick (1.1.3) we arrive at the scalar pressure Poisson equation

$$(3.1.6) \qquad -\Delta p = \mathrm{div}\,(u \cdot \nabla)u,$$

having used $\mathrm{div}\,u = 0$ to kill off the acceleration and diffusion terms. In the following, let us use the term sources for both sources and sinks, so that we need not distinguish $\pm$ signs, especially for terms which may take either sign. For later physical delineation, those with fixed sign will be readily apparent. Since, incompressible or not, we have

$$(3.1.7) \qquad \mathrm{div}\,(u \cdot \nabla)u = \frac{1}{2} \Delta |u|^2 - \omega^2 + (u\omega_y - v\omega_x),$$

where $\omega = \mathrm{curl}\,u = v_x - u_y$ is the vorticity $u = (u, v)$, note that we use u also for the first component of the velocity vector u, the distinction being always evident from the equations, we see that the pressure gradient field sources for (3.1.6) consist of (velocity)^2 potential sources, a vorticity energy, and a vorticity transport. By some algebra and now using $\mathrm{div}\,u = 0$, we arrive from (3.1.7) at a different view, in which all of these pressure gradient sources are combined, namely,

$$(3.1.8) \qquad \Delta p = 2J \begin{pmatrix} u & v \\ x & y \end{pmatrix}$$

where J is the Jacobian of the velocity as a map on the space coordinates. Thus, for "truly" incompressible flow, the total source term for the pressure gradient field evolution is exactly the Jacobian of the evolving velocity. As this measures the change of measure from (x, y) space to (u, v) space, we see that the pressure gradient's role is to vary from exact solenoidality itself only as provoked by the distortion of velocity volume elements.

For not necessarily incompressible flow, we have from $\operatorname{div}(u \cdot \nabla)u$, writing out all terms,

$$
\begin{aligned}
\operatorname{div}(u \cdot \nabla)u &= u_x^2 + u_y^2 + v_x^2 + v_y^2 + u\Delta u_v \Delta v - v_x^2 - u_x^2 + 2v_x u_y \\
&\quad - uu_{yy} - vv_{xx} + uu_{xy} + vu_{xy} \\
&= u_x^2 + v_y^2 + (2u_x v_y - 2u_x v_y) + [uu_{xx} + vv_{yy} + uv_{xy} + vu_{xy}] \\
&\quad + 2v_x u_y \\
&= (\operatorname{div} u)^2 + 2(v_x u_y - u_x v_y) + [u(u_x + v_y)_x + v(v_y + u_x)_y] \\
&= (\operatorname{div} u)^2 - 2J \begin{pmatrix} u & v \\ x & y \end{pmatrix} + u \cdot \nabla(\operatorname{div} u).
\end{aligned}
$$
(3.1.9)

In this third view, that of imperfect incompressibility, we note the appearance of the divergence pressures term and the rotational release factor discussed above. The full pressure Poisson equation from (3.1.5) is now, in place of (3.1.6),

$$
\text{(3.1.10)} \qquad \Delta p = 2J \begin{pmatrix} u & v \\ x & y \end{pmatrix} - (\operatorname{div} u)^2 - u \cdot \nabla(\operatorname{div} u) - (\operatorname{div} u)_t + \frac{1}{\mathrm{Re}} \operatorname{div}(\Delta u).
$$

The pressure gradient is now provoked to depart from exact solenoidality not only by the changing velocity volume distortions but also by the divergence pressures, the work of the rotational release upon the evolving velocity field, the velocity of the divergence pressures, and a last term which would appear to be small for larger Reynolds numbers. In fact, since $\nabla(\operatorname{div} u) = \Delta u + \operatorname{curl}\omega$ for any vector field u, we see that the last term is precisely the divergence of the rotational release, divided by the Reynolds number.

In short, we have thus shown:

$$
\text{(3.1.11)} \quad
\begin{array}{ccccccccccc}
\text{Pressure} & & \text{Velocity} & & \text{Divergence} & & \text{Rotational} & & \text{Divergence} & & \text{Rotational} \\
\text{Gradient} & = & \text{Volume} & + & \text{Pressures} & + & \text{Release} & + & \text{Pressures} & + & \text{Release} \\
\text{Sources} & & \text{Distortions} & & \text{(Squared)} & & \text{(Projected)} & & \text{Velocity} & & \text{Divergence}
\end{array}
$$

The fundamental role of rotational release; its antecedent scalar potential, the divergence pressure; and their consequents, in driving the pressure gradient field, is evident.

Returning to the scheme (3.1.2) with $C = -\Delta$, we see that the preconditioned velocity increment iteration becomes

$$(3.1.12) \qquad B\left(\frac{u^{n+1} - u^n}{\tau}\right) = (1 + (\beta^{-1} - \tau^{-1}))\Delta(u^{n+1} - u^n) + \beta^{-1}\text{curlcurl}\,(u^{n+1} - u^n).$$

In (3.1.12) part of the rotational release has disappeared into the velocity increment diffusion term, the other part (which is a pointwise vorticity entropy) appearing as an additional implicit term. A similar two part effect of the rotational release may be seen in (3.1.3).

It must be admitted that within such implicit schemes for the full Navier–Stokes equations, it is not always easy to foresee the importance of these small scale physical effects of these pressure gradient sources when they occur numerically. There must also be mentioned how these may be affected by the scheme implementation details. For example, an implicit iterative scheme preconditioner of the form

$$(3.1.13) \qquad B = I + (\tau\text{Re}^{-1})^2 R_1 R_2 + \tau\beta^{-1}\nabla A^{-1}\text{div} - \tau\text{Re}^{-1}\Delta$$

may occur, with implementation details such as diffusion $-\Delta = R_1 + R_2$, $R_2 = R_1^* > 0$, pressure Poisson operator $-\text{div}\,\text{grad} = A_1 + A_2$, $A_1 = A_2^* > 0$, $A = (I + \gamma A_1)(I + \gamma A_2)$ occurring in the rotational release term, where R_1, R_2, A_1, A_2 are, for example, triangular matrices. The precise numerical effects of numerical divergence pressures and rotational release should always be carefully examined in terms of the precise implementation details of each such scheme.

REMARK 3.1.1. See Gustafson [6] for further details. It should be emphasized that implicit schemes are of great value for handling higher Reynolds numbers and in some sense cannot be dispensed with altogether. See for example Suito, Ishii, and Kuwahara [1], where flow over an airfoil is computed at Re = 500,000. There the same basic vortex shedding scenario as that reported earlier Gustafson and Leben [4] for much lower Reynolds numbers is observed. The latter code, which allowed no upwinding, will not converge at the higher Reynolds numbers. Therefore a future use of numerical rotational release will be to aid in the design of the most appropriate way to upwind to enable the capture of the appropriate physical solutions.

3.2. Vorticity in the Far Field. Here we clarify the effective far field vorticity boundary condition for the body-centered, body-concentrated grid scheme which was used in the hovering aerodynamics model described in Section 2.2. See Fig. I-5. During the last few years there has been some discussion with others concerning how we obtain such good simulations (excellent vortex detail and lift correspondence with physical experiment), while apparently ignoring the vorticity boundary condition at physical infinity. The quick answer has been that our approach, somewhat novel, maps the physical domain exterior to an airfoil into a near-circular interior domain, which is then mapped to an aspect two computational rectangle with periodic end conditions. See in particular Freymuth, Gustafson, Leben [1, Fig. 6.19, p. 161] or Gustafson and Leben [5, Figs. 3 and 4, pp. 125–126] for the mapping and grid schemes for an arbitrary airfoil. For the hovering motions, an elliptic airfoil shape was used, so the mapping from the exterior physical domain to the interior auxiliary domain is conformal and the auxiliary domain is exactly circular, see Gustafson, Leben, McArthur [1, Fig. 2, p. 49], where however the figures for (b) and (c) should be switched. In our approach, physical infinity is mapped analytically to zero, and is then spread along the left side $\xi = 0$, $0 < \eta < 2$, of our computational rectangle. In this way, we simply avoid having to specify any ad hoc vorticity boundary condition directly in the physical domain, as one needs to do with any direct finite difference or finite element grid placed there. It is extremely convenient for us to simply set the vorticity $\omega = 0$ along the left boundary of our computational rectangle, for otherwise to try to impose some ad hoc but accurate vorticity boundary conditions on some far field physical grid boundary, for these highly translational and highly rotational hovering motions, would be tenuous.

The other half of the quick answer is that we use a body-centered, body-concentrated grid scheme, as shown in Fig. I-5, which gives us high resolution near the airfoil, at the expense of accepting lower resolution in the far field. The preponderance of grid points near the airfoil surface enables good capture of the unsteady vortex shedding events and also enables a corresponding good surface lift computation.

The following is taken from Gustafson [7], in which a less quick, more revealing look at the vorticity in the far field is given. Although the analysis given here is particular to the airfoil motions we are simulating and to our method of discretization, the mode of analysis employed could be useful more generally to provide a better means of understanding the effective vorticity boundary condition at a far field boundary and to enable desired changes to that boundary condition for other specific applications.

In the physical laboratory, the airfoil plunges and pitches in a still air environment, thereby doing work on the fluid. In our computational setup, the fluid environment translates and rotates about the fixed airfoil, thereby doing work on the airfoil. In this situation, the appropriate Navier–Stokes equation becomes

$$\frac{d^2\mathbf{r}}{dt^2} + \frac{d\boldsymbol{\Omega}}{dt} \times \mathbf{r} + \boldsymbol{\Omega} \times \boldsymbol{\Omega} \times \mathbf{r} + 2\boldsymbol{\Omega} \times \mathbf{v} + \frac{d\mathbf{v}}{dt}$$

(3.2.1)   translational eulerian centrifugal coriolis acceleration

$$= -(\mathbf{v} \cdot \nabla)\mathbf{v} + \frac{1}{\text{Re}} \nabla^2 \mathbf{v} - \nabla p.$$

It can be shown that in two dimensions all of these apparent body forces can be taken into an "effective" pressure p, except for the Eulerian term, and that in our hovering motions the latter is a pure curl.

We work with the "apparent" vorticity $\omega^* = \omega_{\text{lab}} + 2\Omega(t)$ where $\Omega(t)$ is a nondimensionalized angular rotation rate of the airfoil. We also work with the perturbation stream function $\psi^* = \psi - \psi_\infty$, where ψ_∞ is the stream function due to the flow at infinity and includes the effects of the angular rotations and translations of the environment about the fixed airfoil. By the use of the noninertial rotating coordinate system and apparent velocities in that system, the Euler term accelerations are accounted for and dropping all asterisks $*$ for convenience, our flow equations then become the streamfunction equation $\nabla^2\psi = -\omega$ and the vorticity transport equation

$$(3.2.2) \quad \frac{\partial\omega}{\partial t} + \frac{1}{h_1 h_2}\left[\frac{\partial}{\partial\xi}(h_2 u\omega) + \frac{\partial}{\partial\eta}(h_1 v\omega)\right] = \frac{L}{R} \cdot \frac{1}{h_1 h_2}\left[\frac{\partial}{\partial\xi}\left(\frac{h_2}{h_1}\frac{\partial\omega}{\partial\xi}\right) + \frac{\partial}{\partial\eta}\left(\frac{h_1}{h_2}\frac{\partial\omega}{\partial\eta}\right)\right]$$

Here u and v are the ξ and η velocity components, L is a characteristic length (related to airfoil chord length), R is the effective (accelerative) Reynolds number, and h_1 and h_2 are grid stretching factors in the ξ (perpendicular to airfoil surface) and η (parallel to airfoil surface) directions, respectively, see Fig. I-5.

We are interested in (3.2.2) in the far field. In the near field, the grid stretching ratio favors h_2, because we concentrate the finite difference grid near the airfoil. Stated another way, we want high aspect ratio grid rectangles, $h_2 \gg h_1$, layered along the airfoil surface. On the other hand, in the far field, this means in our scheme that $h_1 \gg h_2$. There, the radially high aspect ratio grid rectangles elongate out toward infinity. This follows because the distortion factor $f = h_2/h_1$ decreases from its basic airfoil-set aspect ratio, which depends upon the coarseness or fineness of mesh employed there, to zero as the grid tends outward toward physical infinity. This enables a computationally convenient uniform grid in the computational rectangle. To better understand (3.2.2) in the far field, taking into account the grid stretching factors, we now treat u and v as fixed, or more generally we could let them move in bounded intervals, so that we may simplify the analysis. Then the first advection derivative in (3.2.2) becomes

$$(3.2.3) \qquad \frac{\partial}{\partial \xi}(h_2 u \omega) \approx u \omega \frac{\partial}{\partial \xi}(h_2) + u h_2 \frac{\partial}{\partial \xi}(\omega).$$

But in the far field, $h_2 \approx (\eta \cdot \text{some factor})$, and in these radially elongated grid rectangles that factor is relatively unvarying with respect to ξ, so we conclude that the first term of (3.2.3) is $\approx u \omega$ (factor) $\partial \eta / \partial \xi \approx 0$, because the grid has been constructed near-orthogonal. This is the basic idea which makes the following analysis possible.

Thus the accelerative and transport left hand side (LHS) of (3.2.2) becomes, using for efficiency the subscript notation for the derivatives, approximately

$$(3.2.4) \qquad \begin{aligned} &\omega_t + \frac{u\omega}{h_1 h_2}(h_2)_\xi + \frac{u h_2}{h_1 h_2}(\omega)_\xi + \frac{v\omega}{h_1 h_2}(h_1)_\eta + \frac{v h_1}{h_1 h_2}(\omega)_\eta \\ &\approx \omega_t + 0 + \frac{u}{h_1}\omega_\xi + 0 + \frac{v}{h_2}\omega_\eta \end{aligned}$$

and thus

$$(3.2.5) \qquad (\text{LHS}) \approx \omega_t + \frac{v}{h_2}\omega_\eta$$

where the term $u\omega_\xi / h_1$ was dropped because h_1 is so large in the far field. This is the second idea/assumption in our analysis.

The diffusive right hand side (RHS) of (3.2.2) becomes by this analysis

$$(3.2.6) \qquad \begin{aligned} \frac{R}{L} h_1 h_2 (\text{RHS}) &\approx \omega_\xi \frac{\partial}{\partial \xi}\left(\frac{h_2}{h_1}\right) + \frac{h_2}{h_1}\omega_{\xi\xi} + \omega_\eta \frac{\partial}{\partial \eta}\left(\frac{h_1}{h_2}\right) + \frac{h_1}{h_2}\omega_{\eta\eta} \\ &\approx \omega_\xi h_2\left(-\frac{1}{h_1^2}\right) + \frac{h_2}{h_1}\omega_{\xi\xi} + \omega_\eta h_1\left(-\frac{1}{h_2^2}\right) + \frac{h_1}{h_2}\omega_{\eta\eta} \end{aligned}$$

so that

$$(3.2.7) \qquad \begin{aligned} (\text{RHS}) &\approx \frac{L}{R}\left[\frac{\omega_\xi h_2}{h_1 h_2}\left(-\frac{1}{h_1^2}\right) + \frac{h_2}{h_1^2 h_2}\omega_{\xi\xi} + \omega_\eta \frac{h_1}{h_1 h_2}\left(-\frac{1}{h_2^2}\right) + \frac{h_1}{h_2^2 h_1}\omega_{\eta\eta}\right] \\ &\approx \frac{L}{R}\left[\omega_\xi\left(-\frac{1}{h_1^3}\right) + \frac{1}{h_1^2}\omega_{\xi\xi} + \omega_\eta\left(-\frac{1}{h_2^3}\right) + \frac{1}{h_2^2}\omega_{\eta\eta}\right] \end{aligned}$$

and thus

$$(3.2.8) \qquad (\text{RHS}) \approx \frac{L}{R}\left[-\left(\frac{1}{h_2^3}\right)\omega_\eta + \frac{1}{h_2^2}\omega_{\eta\eta}\right]$$

34

where the terms ω_ξ/h_1^3 and $\omega_{\xi\xi}/h_1^2$ were dropped as above because h_1 is so large in the far field. Therefore we may conclude that the effective vorticity boundary condition in the far field is

$$(3.2.9) \qquad \omega_t + \left[\frac{v}{h_2} + \frac{1}{h_2^3}\right]\omega_\eta - \frac{1}{h_2^2}\left(\frac{\text{L}}{\text{R}}\right)\omega_{\eta\eta} = 0$$

This is a one-dimensional (in airfoilwise rotative η spatial direction) advection–diffusion equation in which the far field rectangle grid width h_2 enters in three different powers. Since $v = -h_1^{-1}(\partial\psi/\partial\xi)$ vanishes in the very far field as h_1 becomes very large, we arrive at the simpler equation

$$(3.2.10) \qquad \omega_t + \frac{1}{h_2^3}\omega_\eta - \frac{1}{h_2^2}\left(\frac{\text{L}}{\text{R}}\right)\omega_{\eta\eta} = 0$$

as the simplified far field vorticity boundary equation.

Note that equation (3.2.10) is a grid-dependent, totally (except for the L/R diffusion coefficient, which we regard as a pre-set constant and which we ignore in the present discussion) grid-defined differential equation. Its transport and diffusion coefficients depend only on the size of the η spacings in the orthogonal grid as it has been extended to the far field of the physical domain. This is, to our knowledge, a new or at least unusual way in which one may define the vorticity boundary condition at the far field boundary. Moreover, as one approaches true ∞, for any strictly convex airfoil eventually $h_2 \to \infty$ and the condition (3.2.10) becomes simply $\omega_t = 0$, namely, ω is time-invariant in the far far field. None of this, nor the effective far field vorticity boundary condition (3.2.10), depends on any physically presumed ad hoc boundary condition assigned at a far field outflow boundary, or at physical ∞. A conceptual advantage to our infinite domain method for unsteady flow computations started from time $t = 0$: in any case, in any finite time, one will never reach physical infinity from fluid events all emanating from the airfoil. In this sense, the question of the vorticity boundary condition at ∞ is a moot point.

On the other hand, we have found that the effective vorticity boundary condition occurs for practical purposes at the finite far field boundary of a computational domain (i.e., at the farthest extent of the grid), and is explicitly determined by the grid there. Because most grid generations are global, e.g., adjusted to meet near orthogonality requirements over the whole grid, this means in particular that the spacings h_2 chosen on the airfoil surface will affect the far field vorticity equation (3.2.10). For example, for a slim airfoil, a reasonably coarse uniform h_2 spacing will create far field h_2 directly over the airfoil which are not much changed. This makes equation (3.2.10) of relatively low transport, low diffusion there, hence more manageable numerically. The effective Peclet number is $h_2^{-1}R = O(R)$, the basic Reynolds number of the simulation. Moreover, at the leading and trailing edges, the near-orthogonal grid will expand more rapidly as one moves outward, h_2 will become larger, and the Peclet number will be still smaller. Note however that for very thin airfoils, such as those of the dragonfly, the rapidly expanding airfoil-end grid fan becomes very coarse in the far field. On the other hand, when a fine mesh h_2 spacing is employed, the vorticity boundary equation (3.2.10) becomes highly convective, highly diffusive and less manageable numerically. This probably explains our difficulties in going to finer meshes. When we experimented with 256×256 fine meshes, when we could resolve the flow for Re ~ 1000, the vortex patterns and C_L coefficients were comparable to those found on the 33×33 and 65×65 grids. But for Re ~ 3000 we could not meaningfully resolve the hovering motions on fine grids.

From what we have learned in this analysis, one should feel free to impose a far field vorticity boundary condition resembling (3.2.10) but if need be always modified to be amenable to the overall solution of the oscillatory problem at hand. For example early in our investigations we went to far field line relaxations (for the stream function in $\nabla^2\psi = -\omega$) in order to resolve the hovering motions as Reynolds number increased. In retrospect, this is an indirect form of modifying (through the velocity v obtained from the stream function ψ) the vorticity equation (3.2.9) in order to resolve the whole discretized Navier–Stokes system. Closer consideration of the equivalent of (3.2.9) or (3.2.10) for other discretization methods could lead to useful improvements in setting a far field discrete vorticity boundary condition appropriate to other problems in which the modelling of far field vorticity is an important consideration. For example, in spite of its minor physical importance to the resolution of the near field vortex shedding events which dominate the aerodynamics of our hovering motions, it is in the far field that we encounter the most severe numerical limitations. Then the actual discretizations which have been employed for the partial derivatives of the model, e.g., in our case, $\omega_t, \omega_\eta,$ and $\omega_{\eta\eta}$ within an ADI finite difference scheme, should be placed into (3.2.10) or its equivalent, to enable a better understanding and possible correction of effects, e.g., false numerical oscillations, which may be occurring in the far field. Both for its own interest and for illustration, let us carry out this recommended procedure for our far field vorticity condition (3.2.10). With explicit time differencing and upwind convective differencing and centered diffusion differencing, from (3.2.10) we arrive at

(3.2.11)

$$\omega(\eta, t + \Delta t) = \frac{\Delta t}{h_2^4}\left(\frac{L}{R}\right)\omega(\eta + h_2) + \left(1 - \frac{\Delta t}{h_2^4}\left(1 + 2\left(\frac{L}{R}\right)\right)\right)\omega_\eta + \frac{\Delta t}{h_2^4}\left(1 + \left(\frac{L}{R}\right)\right)\omega(\eta - h_2)$$

This is a convex combination of $\omega(\eta + h_2)$, ω_η, and $\omega(\eta - h_2)$ and hence by standard considerations will guarantee the stability of the scheme, provided that

(3.2.12)
$$\Delta t < \frac{h_2^4}{1 + 2\left(\frac{L}{R}\right)}.$$

For relatively large Reynolds numbers R, we now understand, in terms of our far field grid spacings, how small we must take Δt, e.g., $\Delta t < h_2^4$, to keep vorticity bounded in the far field.

3.3. Splitting Errors. Discrete operator splitting was necessary in the development of efficient CFD schemes. For example, significant time-saving advantages accrue if one can force the algebra in the linear solvers to tridiagonal matrix forms. Operator splitting techniques go under various names such as time splitting, method of fractional steps, approximate factorization methods, projection methods, alternating–direction–implicit schemes, and so on. Important names are attached: Samarskii, Chorin, Yanenko, Peaceman, Rachford, and others. The following is taken from Gustafson and McArthur [1]. Our point is that all of these methods create a symmetry breaking and therefore a solution selection which may or may not be correct. A second point is that for higher resolution CFD, especially for the computation of time dependent oscillating solutions in which one needs to compute correct physical oscillations while rejecting spurious numerical oscillations, a reexamination of splitting methods will be needed. See Gustafson and McArthur [2] for further results. Here we shall be content to illustrate these points by means of a simplified example.

Let us take the point of view that numerical schemes are exponential approximations. Thus the famous Crank–Nicolson scheme for the heat equation $w_t = w_{xx} \equiv -Aw$ becomes

(3.3.1)
$$\omega^{n+1} = e^{-\frac{t}{2}A}e^{-\frac{t}{2}A}\omega^n = \left(I + \frac{t}{2}A\right)^{-1}\left(I - \frac{t}{2}A\right)\omega^n$$

where one first truncates the series $e^{-\frac{t}{2}A} = I - \frac{t}{2}A + \left(\frac{t}{2}A\right)^2/2! + \cdots$ and then truncates the inverse series $e^{-\frac{t}{2}A} = 1/\left(I + \frac{t}{2}A + \left(\frac{t}{2}A\right)^2/2! + \cdots\right)$. Taking the one-dimensional Navier–

Stokes model problem

$$(3.3.2) \qquad \omega_t = -u\omega_x + k\omega_{xx} = -uB\omega + kA\omega = C\omega$$

in which we regard the velocity u and viscosity k as fixed, by the exponential theory we expect a solution

$$\omega(t) = e^{tC}\omega(0) = e^{t(-uB+kA)}\omega(0)$$

Let us consider the effects of exponential operator splitting here, using a centered diffusion stencil and a centered advection stencil. Then locally on a grid of size $\Delta x = h$ we have the representations

$$(3.3.3) \qquad C = -uB + kA = \frac{-u}{2h}\begin{bmatrix} 0 & 1 & 0 \\ -1 & 0 & 1 \\ 0 & -1 & 0 \end{bmatrix} + \frac{k}{h^2}\begin{bmatrix} -2 & 1 & 0 \\ 1 & -2 & 1 \\ 0 & 1 & -2 \end{bmatrix}$$

An eigenvalue computation (using MAPLE) gives the eigenvalues of C

$$(3.3.4) \qquad \frac{-2k}{h^2}, \quad \frac{-2k}{h^2} \pm \frac{\sqrt{2(4k^2 - u^2h^2)}}{2h^2}$$

and the exponential

$$(3.3.5)$$
$$e^{tC} = \frac{1}{2} e^{-\frac{2kt}{h^2}} \begin{bmatrix} 1 + \cosh\left(\frac{Gt}{2h^2}\right) & \frac{G}{2k+uh}\sinh\left(\frac{Gt}{2h^2}\right) & \frac{-uh+2k}{uh+2k}\left(-1+\cosh\left(\frac{Gt}{2h^2}\right)\right) \\ \frac{G}{2k-uh}\sinh\left(\frac{Gt}{2h^2}\right) & 2\cosh\left(\frac{Gt}{2h^2}\right) & \frac{G}{2k+uh}\sinh\left(\frac{Gt}{2h^2}\right) \\ \frac{uh+2k}{-uh+2k}\left(1-\cosh\left(\frac{Gt}{2h^2}\right)\right) & \frac{G}{2k-uh}\sinh\left(\frac{Gt}{2h^2}\right) & 1+\cosh\left(\frac{Gt}{2h^2}\right) \end{bmatrix}$$

where $G = \sqrt{2(4k^2 - u^2h^2)}$. When G is real, there are no oscillatory terms in the matrix M above. But when G becomes imaginary, all hyperbolic trigonometric terms become trigonometric, e.g., $\sinh ix = i\sin x$, $\cosh ix = i\sin x$. Therefore numerical oscillations begin only when $|u| \geqq 2k/h$. The local velocity exceeds the ability of the grid to transmit information.

On the other hand, a similar analysis (using MAPLE) reveals that

$$(3.3.6) \qquad e^{tkA} = \frac{1}{2} e^{-\frac{2kt}{h^2}} \begin{bmatrix} 1+\cosh\left(\frac{\sqrt{2}kt}{h^2}\right) & \sqrt{2}\sinh\left(\frac{\sqrt{2}kt}{h^2}\right) & -1+\cosh\left(\frac{\sqrt{2}kt}{h^2}\right) \\ \sqrt{2}\sinh\left(\frac{\sqrt{2}kt}{h^2}\right) & 2\cosh\left(\frac{\sqrt{2}kt}{h^2}\right) & \sqrt{2}\sinh\left(\frac{\sqrt{2}kt}{h^2}\right) \\ -1+\cosh\left(\frac{\sqrt{2}kt}{h^2}\right) & \sqrt{2}\sinh\left(\frac{\sqrt{2}kt}{h^2}\right) & 1+\cosh\left(\frac{\sqrt{2}kt}{h^2}\right) \end{bmatrix}$$

which is never oscillatory, and

$$(3.3.7) \qquad e^{-tuB} = \frac{1}{2} \begin{bmatrix} 1+\cos\left(\frac{\sqrt{2}ut}{2h}\right) & -\sqrt{2}\sin\left(\frac{\sqrt{2}ut}{2h}\right) & 1-\cos\left(\frac{\sqrt{2}ut}{2h}\right) \\ \sqrt{2}\sin\left(\frac{\sqrt{2}ut}{2h}\right) & 2\cos\left(\frac{\sqrt{2}ut}{2h}\right) & -\sqrt{2}\sin\left(\frac{\sqrt{2}ut}{2h}\right) \\ 1-\cos\left(\frac{\sqrt{2}ut}{2h}\right) & \sqrt{2}\sin\left(\frac{\sqrt{2}ut}{2h}\right) & 1+\cos\left(\frac{\sqrt{2}ut}{2h}\right) \end{bmatrix}$$

which is always oscillatory. Thus this exponential operator symmetry breaking 'creates' an imaginary part of the solution no matter what the parameter values.

A remedy for such unwanted numerical oscillation is upwinding, in which one replaces a second order discretization by accepting a first order discretization with preferred (upwind) direction. In the above example it may be seen that for the symmetry breaking, the evolving solution is undervalued. Also, in the two space dimensions hovering experiments, the velocity u components rapidly change sign in the highly rotational flow. To better understand splitting errors we used the Lie theory in which for a given e^X and e^Y we are guaranteed an operator Z such that the exponentially split $e^X e^Y = e^Z$ where

$$(3.3.8) \qquad Z = X + Y + \frac{1}{2}\,[X,Y] + \frac{1}{12}\,[X,[X,Y]] + \frac{1}{12}\,[Y,[Y,X]] - \cdots$$

It is instructive to see what this means for the above model problem, in which $X = -uB$ and $Y = kA$. Then

$$(3.3.9) \qquad [X,Y] = -\frac{ku}{h^3}\begin{bmatrix} -1 & 0 & 0 \\ 0 & 0 & 0 \\ 0 & 0 & 1 \end{bmatrix}$$

splits into forward and backward 'characteristic' directions, the finer the grid the worse this splitting, and

$$(3.3.10) \qquad \frac{1}{12}\,[X,[X,Y]] + \frac{1}{12}\,[Y,[Y,X]] = \begin{bmatrix} 0 & uh+2k & 0 \\ uh-2k & 0 & uh+2k \\ 0 & uh-2k & 0 \end{bmatrix}$$

enables an exact error analysis. A more detailed theory of Lie corrections for general CFD schemes will be given elsewhere, Gustafson and McArthur [2].

To summarize this brief account, we may state

PROPOSITION 3.3.1. Numerical splitting errors may be analyzed by stencil exponentiation, from which Lie theory provides in principle exact error analysis and appropriate correction terms for any scheme.

REMARK 3.3.2. It would be interesting to extend all considerations of this chapter to three dimensions and toward eventual turbulence studies.

Comments and Bibliography

There are roughly three ways to discretize partial differential equations: finite difference methods, finite element methods, finite spectral methods. See the discussion in Gustafson [1]. There are strong arguments (and strong adherents) for each method. For example it can be claimed that FDM better represents the differential equations, FEM better represents irregular boundaries, FSM better enables higher resolution. In all approaches the effects of the discretization on the resulting linear matrix equations must be kept in mind because the latter will consume the bulk of the computation time. All three methods for example have been applied to the cavity flow problem of Chapter 1.

Concerning Chapter 2, there is a large literature already on computational aerodynamics and we make no attempt to present it here. For comparison to the approach of Section 2.2, see for example Osswald, Ghia and Ghia [1], which is similar.

The notion of numerical rotational release introduced in Chapter 3, and the issue of how upwinding may distort a representation of a physical flow, although clearly related for simulations of two- and three-dimensional incompressible fluid flows, should be regarded more generally as distinct features of a computational scheme. The former addresses fluid circulation properties whereas the latter usually addresses fluid convection properties. Of course through the Navier–Stokes equations, these properties are connected in any discretization.

40

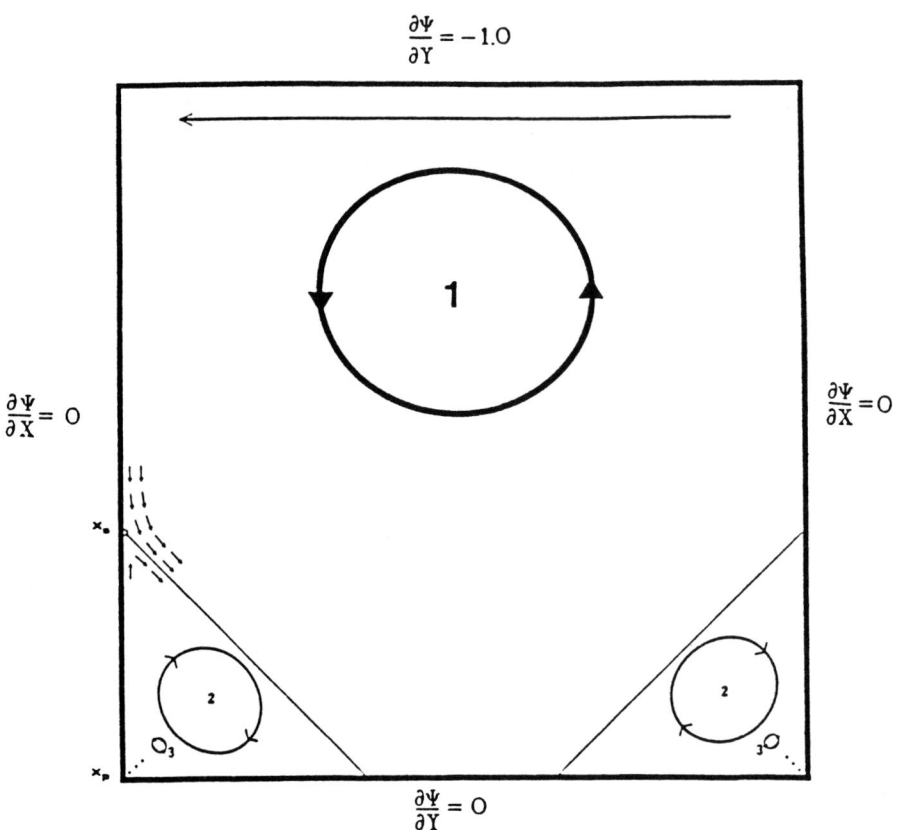

Figure 1. Driven Cavity Flow, A = 1.

1:   Principal Vortex
2:   Secondary Corner Vortex
3:   Tertiary Corner Vortex
x_s: Separation Point
x_p: Provocation point

Fig. I-1. Driven cavity flow, aspect ratio $A = 1$.
From Gustafson and Leben [1].

Table 1: Local Maximum Stream Function Intensities and Associated Residual and Iterations [2].

Vortex	Intensity	Residual	Iteration
ψ_1	0.996×10^{-1}	0.88×10^{-5}	53
ψ_2	-2.29×10^{-6}	0.97×10^{-5}	10
ψ_3	6.55×10^{-11}	0.90×10^{-5}	16
ψ_4	-1.87×10^{-15}	0.94×10^{-5}	21
ψ_5	5.33×10^{-20}	0.95×10^{-5}	26
ψ_6	-1.52×10^{-24}	0.95×10^{-5}	31
ψ_7	4.34×10^{-29}	0.98×10^{-5}	36
ψ_8	-1.24×10^{-33}	0.98×10^{-5}	41
ψ_9	3.53×10^{-38}	0.97×10^{-5}	46
ψ_{10}	-1.01×10^{-42}	0.94×10^{-5}	51
ψ_{11}	2.88×10^{-47}	0.88×10^{-5}	56
ψ_{12}	-8.19×10^{-52}	0.95×10^{-5}	60
ψ_{13}	2.34×10^{-56}	0.85×10^{-5}	65
ψ_{14}	-6.69×10^{-61}	0.85×10^{-5}	70
ψ_{15}	1.91×10^{-65}	0.95×10^{-5}	74
ψ_{16}	-5.45×10^{-70}	0.15×10^{-4}	100
ψ_{17}	1.55×10^{-74}	0.30×10^{-4}	100
ψ_{18}	-4.42×10^{-79}	0.61×10^{-4}	100
ψ_{19}	1.27×10^{-83}	0.12×10^{-3}	100
ψ_{20}	-3.61×10^{-88}	0.24×10^{-3}	100
ψ_{21}	1.03×10^{-92}	0.48×10^{-3}	100
ψ_{22}	-2.94×10^{-97}	0.19×10^{-2}	100
ψ_{23}	8.39×10^{-102}	0.19×10^{-2}	100
ψ_{24}	-2.40×10^{-105}	0.78×10^{-2}	100
ψ_{25}	6.83×10^{-111}	0.15×10^{-1}	100
ψ_{26}	-1.95×10^{-115}	0.31×10^{-1}	100

Table I-2. Cavity corner stream function intensities. From Gustafson and Leben [3].

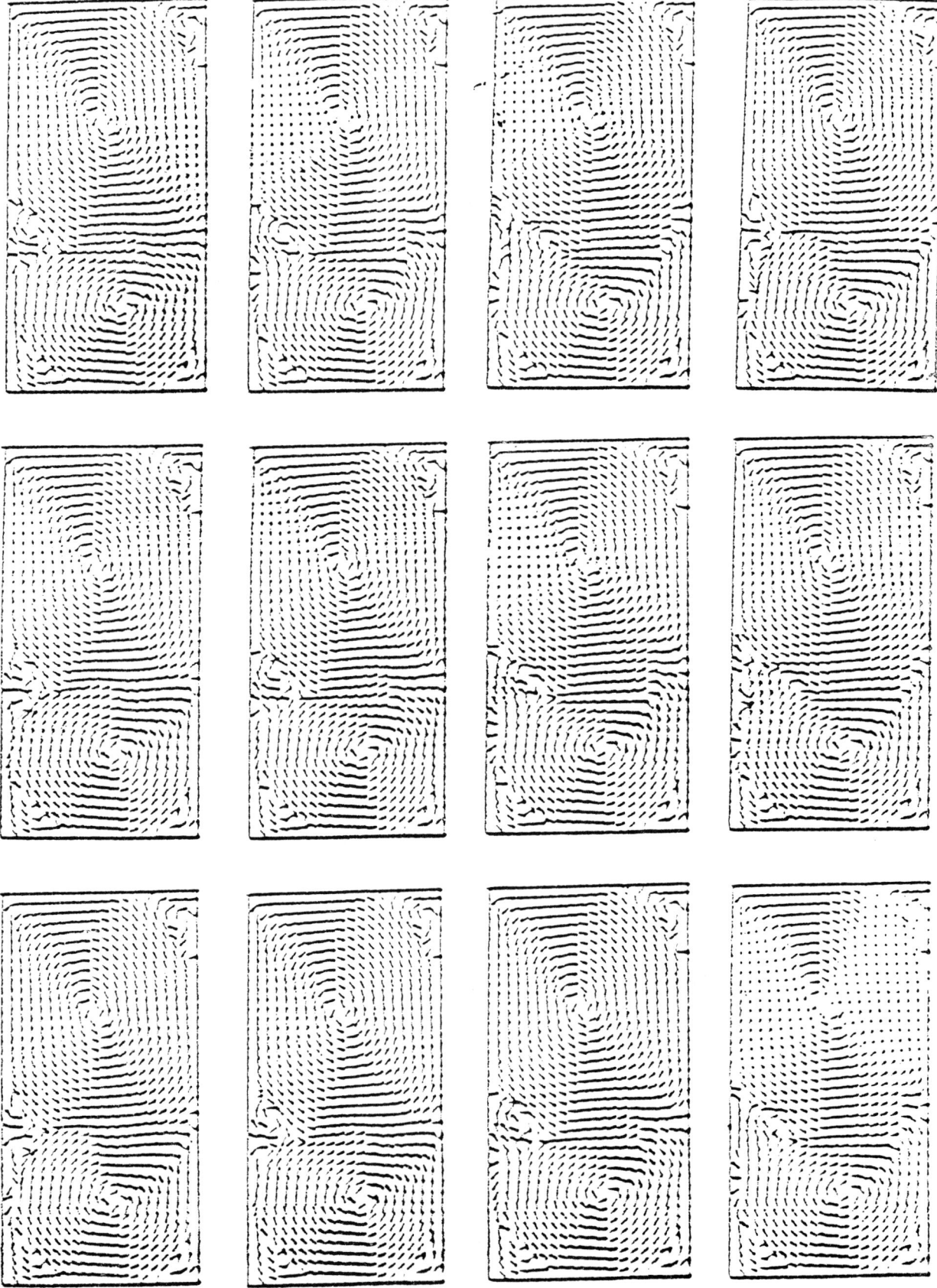

Fig. I-3. Cavity Hopf bifurcation, $A = 2$, Re $= 10,000$.
From Gustafson and Halasi [2].

Mode 1 Hovering

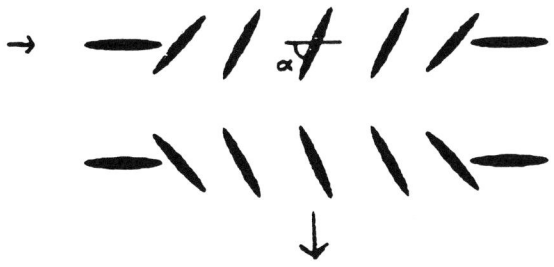

Mode 2 Hovering

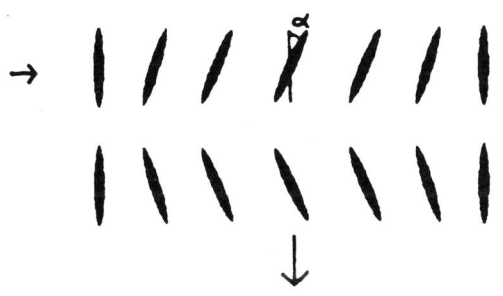

Mode 3 Hovering

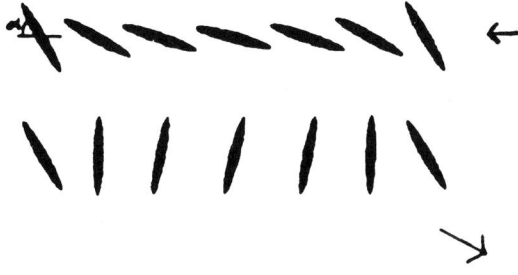

Fig. I-4. Three hovering modes combining plunge and pitch.
From Gustafson, Leben, McArthur [1].

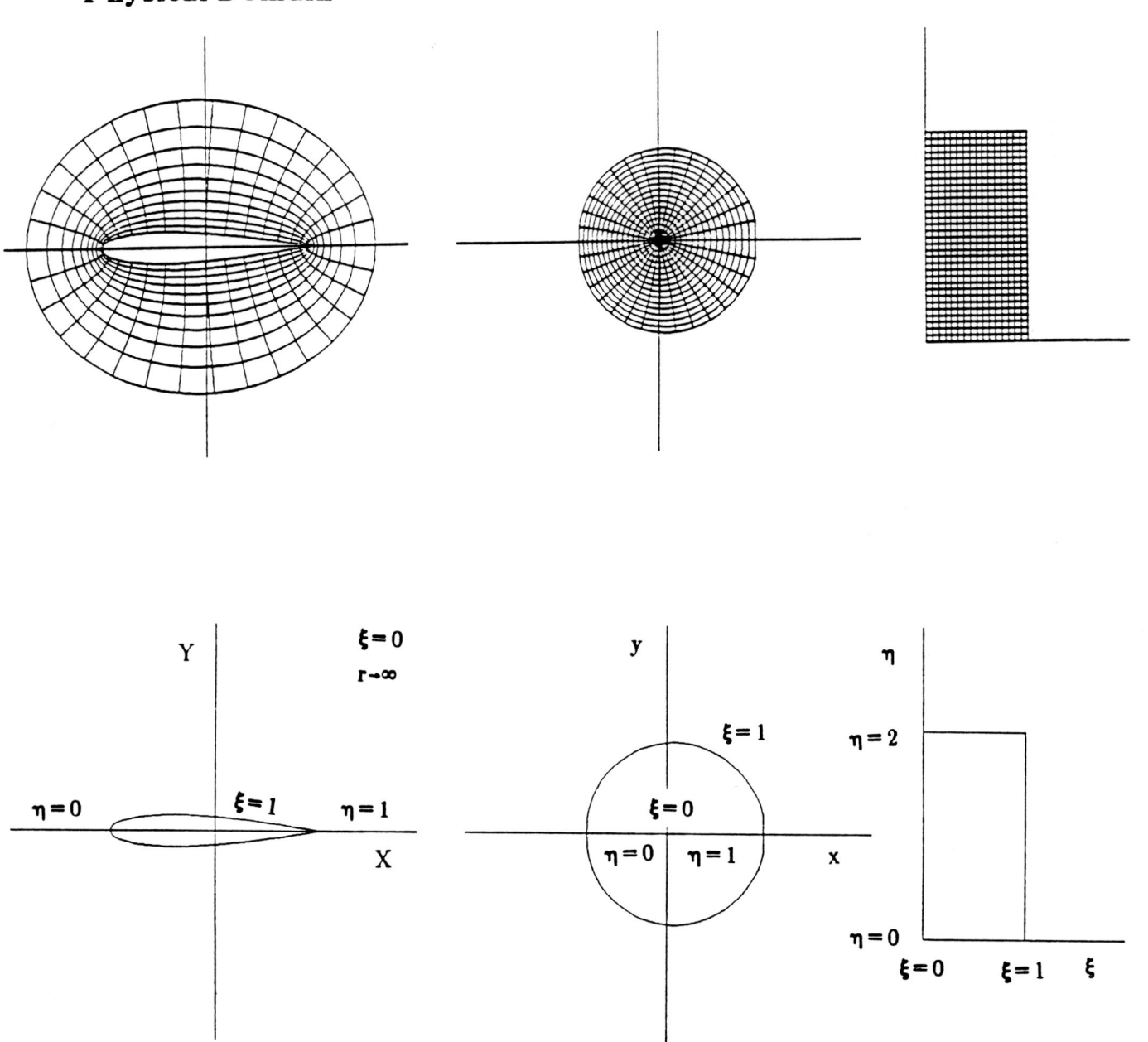

Fig. I-5. Airfoil domains and coordinate systems.
From Gustafson and Leben [4].

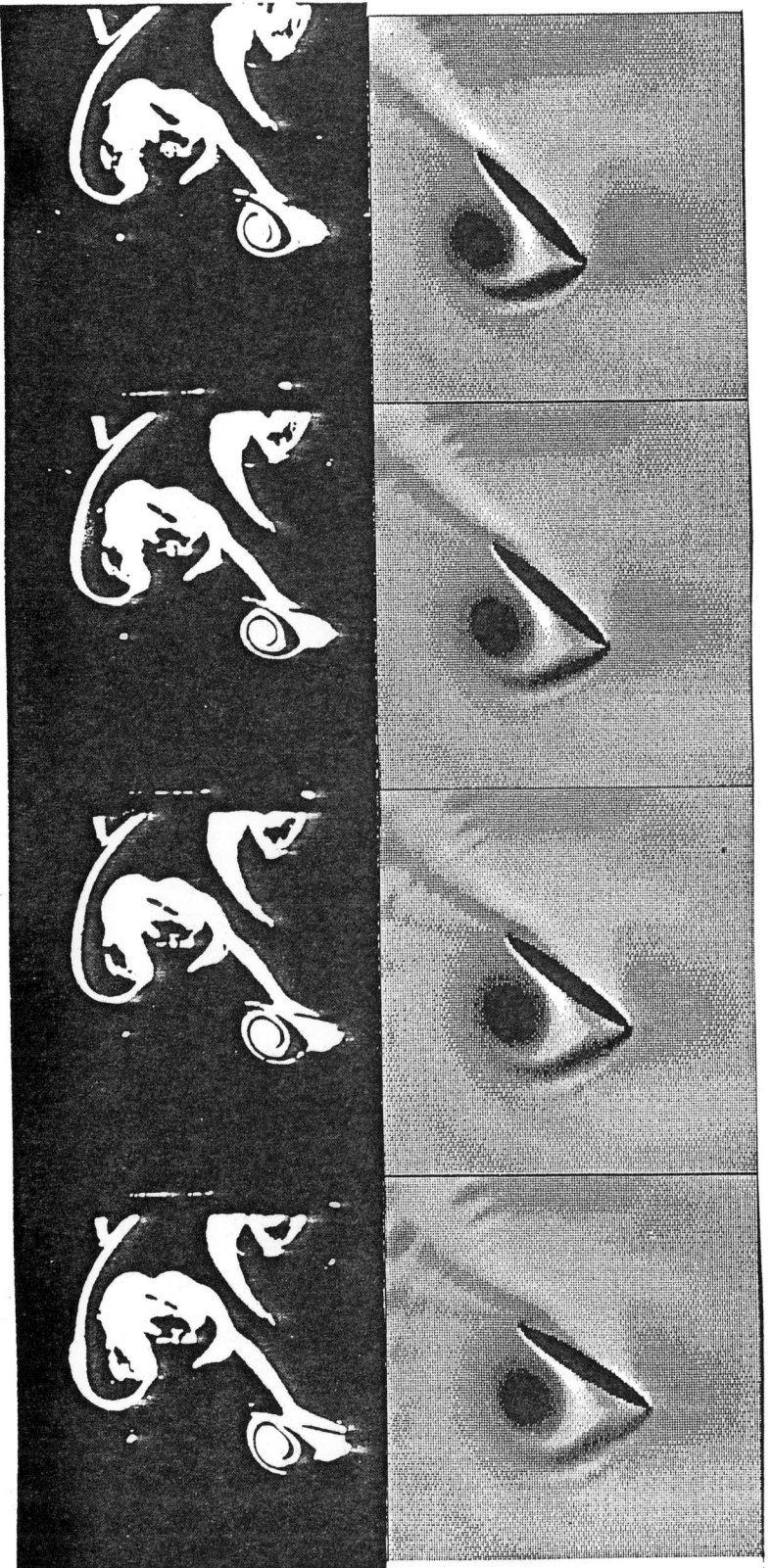

Mode 1

$$\alpha_a = 66°$$
$$h_{a/c} = 1.5$$
$$f = 1.0 \text{ Hz}$$
$$R_f = 340$$
$$\Delta t = 1/64 \text{ second}$$

Experimental sequence on left is slightly askew to view airfoil. Note the very good correlation of the computer simulation on the right.

Fig. I-6. Hovering aerodynamics, experimentally and computationally. From Gustafson and Leben [5].

46

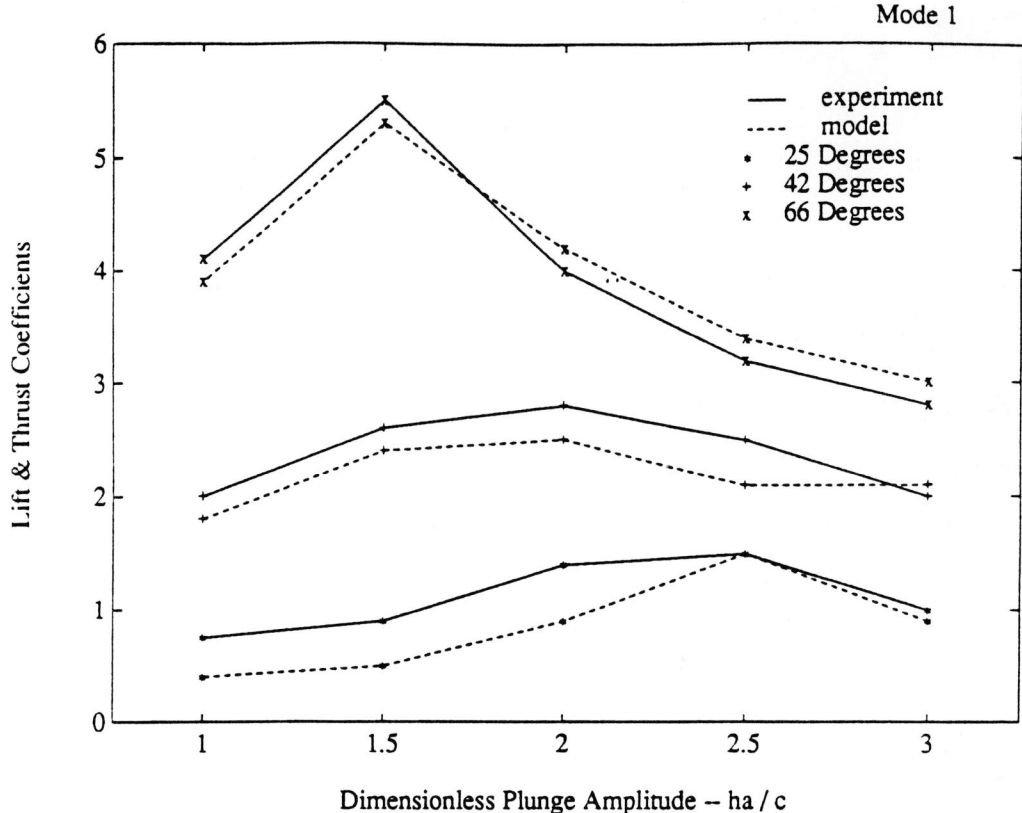

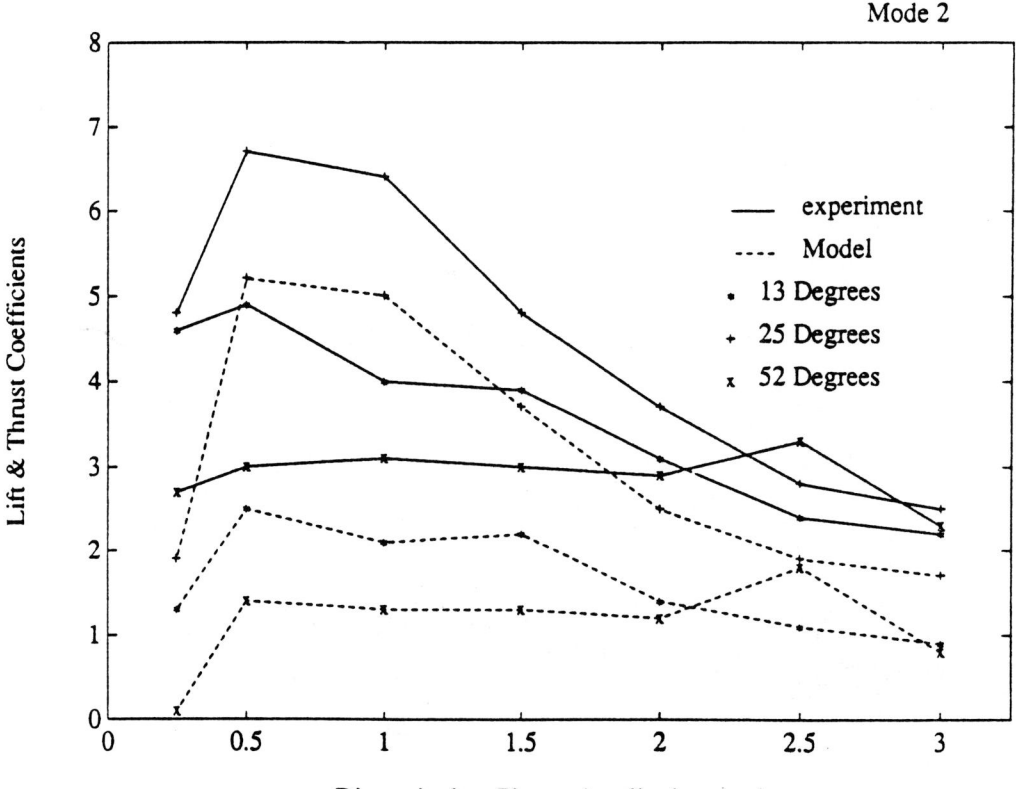

Fig. I-7. Hovering lift C_L and thrust C_T comparisons.
From Gustafson, Leben, McArthur [1].

References

A. ACRIVOS AND F. PAN

[1] Steady flows in rectangular cavities, *J. Fluid Mech.* **28** (1967), 643–655.

A. AZUMA

[1] *The Biokinetics of Flying and Swimming*, Springer, Tokyo (1992).

A. AZUMA AND T. WATANABE

[1] Flight performance of a dragonfly, *J. Exp. Biology* **137** (1988), 221–252.

G. BATCHELOR

[1] *An Introduction to Fluid Dynamics*, Cambridge (1967).

A. BENJAMIN AND V. DENNY

[1] On the convergence of numerical solutions for 2-D flows in a cavity at large Re, *J. Comp. Phys.* **33** (1979), 340–358.

S. CHAN

See P. Gresho, S. Chan, R. Lee and C. Upson.

W. DEAN AND P. MONTAGNON

[1] On the steady motion of viscous liquid in a corner, *Proc. Cambridge Phil. Soc.* **45** (1949), 389–394.

V. DENNY

See A. Benjamin and V. Denny.

P. DUCK

[1] Oscillatory flow inside a square cavity, *J. Fluid Mech.* **122** (1982), 215–234.

A. FORTIN, M. JARDAK, J. GERVAIS AND R. PIERRE

[1] Localization of Hopf bifurcations in fluid flow problems, *Intern. J. Numer. Meths. Fluids*, to appear.

P. FREYMUTH

[1] Propulsive vortical signature of plunging and pitching airfoils, *AIAA J.* **26** (1988), 881–883.

[2] Thrust generation by an airfoil in hover modes, *Experiments in Fluids* **9** (1990), 17–24.

P. FREYMUTH, K. GUSTAFSON AND R. LEBEN

[1] Visualization and computation of hovering mode vortex dynamics, in *Vortex Methods and Vortex Motion*, (K. Gustafson, J. Sethian, eds.) SIAM, Philadelphia (1990), 143–169.

D. FU

See H. Liu, D. Fu and Y. Ma.

L. FUCHS AND N. TILLMARK

[1] Numerical and experimental study of driven flow in a polar cavity, *Intern. J. Numer. Meths. Fluids* **5** (1985), 311–329.

J. GERVAIS

See A. Fortin, M. Jardak, J. Gervais and R. Pierre.

K. GHIA

See U. Ghia, K. Ghia and C. Shin; G. Osswald, K. Ghia and U. Ghia.

U. GHIA, K. GHIA AND C. SHIN

[1] High-Re solutions for incompressible flow using the Navier–Stokes equations and a multi-grid method, *J. Comp. Phys.* **48** (1982), 387–411.

See also G. Osswald, K. Ghia and U. Ghia.

R. GLOWINSKI, H. KELLER AND L. REINHART

[1] Continuation-conjugate gradient methods for the least squares solution of nonlinear boundary value problems, *SIAM J. Sci Statist. Comp.* **6** (1985), 793–832.

J. GOODRICH, K. GUSTAFSON AND K. HALASI

[1] Hopf bifurcation in driven cavity flow, *J. Comp. Phys.* **90** (1990), 219–261.

P. GRESHO, S. CHAN, R. LEE AND C. UPSON

[1] A modified finite element method for solving the time-dependent, incompressible Navier–Stokes equations. Part 2: Applications, *Intern. J. Numer. Meths. Fluids* **4** (1984), 619–640.

P. GRISVARD

[1] *Elliptic Problems in Nonsmooth Domains*, Pittman, London (1985).

K. GUSTAFSON

[1] *Introduction to Partial Differential Equations*, Wiley (1980, 1987), Kaigai, Tokyo (1991, 1992), International Journal Services, Calcutta (1993).

[2] Vortex Separation and fine structure dynamics, *Appl. Num. Math.* **3** (1987), 167–182.

[3] Four principles of vortex motion, in *Vortex Methods and Vortex Motion* (K. Gustafson, J. Sethian, eds.), SIAM, Philadelphia (1990), 95–141.

[4] Theory and computation of periodic solutions of autonomous partial differential equation boundary value problems, with application to the driven cavity problem, *Mathl. Comput. Modelling* **22** (1995), 57–76.

[5] Biological dynamical subsystems of hovering flight, *Mathematics and Computers in Simulation* **40** (1996), 397–410.

[6] A physical interpretation of Kobelkov's fully implicit term, to appear.

[7] Vorticity in the far field, to appear.

See also P. Freymuth, K. Gustafson and R. Leben; J. Goodrich, K. Gustafson and K. Halasi.

K. GUSTAFSON AND K. HALASI

[1] Vortex dynamics of cavity flows, *J. Comp. Phys.* **64** (1986), 279–319.

[2] Cavity flow dynamics at higher Reynolds number and higher aspect ratio, *J. Comp. Phys.* **70** (1987), 271–283.

K. GUSTAFSON AND R. LEBEN

[1] Multigrid calculation of subvortices, *Appl. Math. Comput.* **19** (1986), 89–102.

[2] Vortex subdomains, in *Proc. First Intern. Symp. on Domain Decomp. Methods for Partial Differential Equations* (R. Glowinski, G. Golub, G. Meurant, J. Periaux, eds.), SIAM, Philadelphia (1988), 370–380.

[3] Multigrid localization and multigrid grid generation for the computation of vortex structures and dynamics of flows in cavities and about airfoils, in *Multigrid Methods* (S. McCormick, ed.), Marcel Dekker, New York (1988), 229–249.

[4] Robust multigrid computation and visualization of separation and vortex evolution in aerodynamic flows, in *Proc. 1st Nat. Fl. Dyns. Congress* (K. Ghia, ed.), Cincinnati, Ohio, AIAA/ASME/SIAM/APS (1988), 174–184. See also AIAA Paper 88–3604–CP.

[5] Computation of dragonfly aerodynamics, *Computer Physics Communications* **65** (1991), 121–132.

K. GUSTAFSON, R. LEBEN AND J. MCARTHUR

[1] Lift and thrust generation by an airfoil in hover modes, *Computational Fluid Dynamics*

J. **1** (1992), 47–57.

K. GUSTAFSON, R. LEBEN, J. MCARTHUR AND M. MUNDT

[1] Qualitative features of high lift hovering dynamics and inertial manifolds, *Theor. and Comput. Fl. Dyns.* **8** (1996), 89–104.

K. GUSTAFSON AND J. MCARTHUR

[1] Symmetry breaking in CFD: causes and consequences, *Inter. J. Modern Phys. C* **5** (1994), 189–194.

[2] Splitting errors in CFD, to appear.

K. GUSTAFSON AND J. SETHIAN

[1] *Vortex Methods and Vortex Motion*, SIAM, Philadelphia (1990).

K. HALASI

See J. Goodrich, K. Gustafson and K. Halasi; K. Gustafson and K. Halasi.

K. ISHII

See H. Suito, K. Ishii and K. Kuwahara.

M. JARDAK

See A. Fortin, M. Jardak, J. Gervais and R. Pierre.

M. KAWAGUTI

[1] Numerical solution of the Navier–Stokes equations for the flow in a two-dimensional cavity, *J. Phys. Soc. Japan* **16** (1961), 2307–2315.

H. KELLER

See R. Glowinski, H. Keller and L. Reinhart; R. Schreiber and H. Keller.

G. KOBELKOV

[1] Solution of the problem of steady state free convection, *Dokl. Akad. Nauk.* SSSR **255** (1980); English transl. in *Soviet Math. Dokl.* **22** (1980), 676–681.

K. KUWAHARA

See H. Suito, K. Ishii and K. Kuwahara.

O. LADYZHENSKAYA

[1] *The Mathematical Theory of Viscous Incompressible Flow*, Gordon and Breach, New York (1963).

R. LEBEN

    See P. Freymuth, K. Gustafson and R. Leben; K. Gustafson and R. Leben; K. Gustafson, R. Leben and J. McArthur; K. Gutafson, R. Leben, J. McArthur and M. Mundt.

R. LEE

    See P. Gresho, S. Chan, R. Lee and C. Upson.

H. LIU, D. FU AND Y. MA

    [1] Upwind compact method and direct numerical simulation of driven flow in a square cavity, *Science in China* **36** (1993), 1347–1357.

H. LUGT

    [1] *Vortex Flow in Nature and Technology*, Wiley, New York (1983).

M. LUTTGES

    See S. Somps and M. Luttges.

Y. MA

    See H. Liu, D. Fu and Y. Ma.

J. MARSDEN AND M. MCCRACKEN

    [1] *The Hopf Bifurcation and Its Applications*, Springer, New York (1976).

J. MCARTHUR

    See K. Gustafson, R. Leben and J. McArthur; K. Gustafson, R. Leben, J. McArthur and M. Mundt; K. Gustafson and J. McArthur.

M. MCCRACKEN

    See J. Marsden and M. McCracken.

H. MOFFATT

    [1] Viscous and resistive eddies near a sharp corner, *J. Fluid Mech.* **18** (1964), 1–18.

    [2] Viscous eddies near a sharp corner, *Arch. Mech. Stosowanej* **16** (1964), 365–372.

P. MONTAGNON

    See W. Dean and P. Montagnon.

M. MUNDT

    See K. Gustafson, R. Leben, J. McArthur and M. Mundt.

52

G. OSSWALD, K. GHIA AND U. GHIA

    [1] Simulation of dynamic stall phenomenon using unsteady Navier–Stokes equations, *Computer Physics Communications* 65 (1991), 209–218.

F. PAN

    See A. Acrivos and F. Pan

R. PEYRET AND T. TAYLOR

    [1] *Computational Methods for Fluid Flow*, Springer, Berlin (1983).

R. PIERRE

    See A. Fortin, M. Jardak, J. Gervais and R. Pierre.

LORD RAYLEIGH

    [1] On the question of the stability of the flow of fluids, *Phil. Magazine* **34** (1892), 59–70.

    [2] Hydrodynamical notes, *Phil. Magazine* **21** (1911), 177–195.

L. REINHART

    See R. Glowinski, H. Keller and L. Reinhart.

R. SCHREIBER AND H. KELLER

    [1] Driven cavity flows by efficient numerical techniques, *J. Comp. Phys.* **49** (1983), 310–333.

D. SERRE

    [1] Equations de Navier–Stokes stationaires avec données peu reguliere, *Ann. Scuola Norm. Sup. Pisa Cl. Sci.* **4** (1984), 543–559.

J. SETHIAN

    See K. Gustafson and J. Sethian.

J. SHEN

    [1] Hopf bifurcation of the unsteady regularized driven cavity, *J. Comp. Phys.* **95** (1991), 228–245.

C. SHIN

    See U. Ghia, K. Ghia and C. Shin.

S. SOMPS AND M. LUTTGES

    [1] Dragonfly flight: novel uses of unsteady separated flows, *Science* **228** (1985), 1326–1329.

H. SUITO, K. ISHII AND K. KUWAHARA

[1] Simulation of dynamic stall by multi-directional finite difference method, *26th AIAA Fluid Dynamics Conference*, June 19–22, 1995, San Diego, CA, AIAA Paper 95–2264.

S. TANEDA

[1] Visualization of separating Stokes flows, *J. Phys. Soc. Japan* **46** (1979), 1935–1942.

T. TAYLOR

See R. Peyret and T. Taylor.

R. TEMAM

[1] *Navier–Stokes Equations*, North Holland, Amsterdam (1985).

[2] *Navier–Stokes Equations and Nonlinear Functional Analysis*, SIAM, Philadelphia (1983).

N. TILLMARK

See L. Fuchs and N. Tillmark.

C. UPSON

See P. Gresho, S. Chan, R. Lee and C. Upson.

T. WATANABE

See A. Azuma and T. Watanabe.

K. WINTERS

[1] Bifurcation and stability: a computational approach, *Computer Phys. Communications* **65** (1991), 299–309.

PART II

Recent Developments in Mathematical Physics

Perspective. In this second part, we present new developments in three currently important directions in mathematical physics. The first chapter is concerned with one of the fundamental questions of physics: the meaning and the role of irreversibility. The second chapter establishes new connections between the theory of wavelets and that of stochastic processes. The third chapter treats chaos as seen through iterative dynamical systems. Although these three directions in mathematical physics are developing independently, there are common threads among them.

Chapter 1 treats the theory of I. Prigogine, in which a central question is how to reconcile deterministic and stochastic descriptions. The underlying dynamical systems are shown to be necessarily Kolmogorov flows. New views about irreversibility, ordering of time, and second laws of nature are presented. These developments commence around 1980 and are brought up to date to current research.

Chapter 2 presents new connections between Kolmogorov systems and wavelets. In particular, it is shown that any wavelet multiresolution analysis possesses a Time operator from the theory of statistical physics. Moreover, links to martingales and stochastic processes are established.

Chapter 3 presents a new 'gap' theory for the investigation of the fractal (information) dimension of iterated maps occurring in the ergodic theory of chaos. Additionally a fundamental new understanding of neural learning dynamics is presented. These are also current (1995, 1996) research results.

Chapter 1   Probabilistic and Deterministic Description

1.1. Theories of Prigogine and Kolmogorov. The results presented in this chapter originated in 1974 in a collaboration Gustafson and Misra [1], Gustafson [1], Misra and Sudershan [1], concerned with models of decay of unstable quantum mechanical particles, and continued in collaboration with the Prigogine, et al., Brussels school in connection with mathematical models of irreversibility in statistical physics. In particular, in four basic papers Misra, Prigogine, and Courbage [1], [2], Misra and Prigogine [1], Goldstein, Misra, and Courbage [1] in 1979–1981, the results stated below in Theorems 1.1.5 and 1.1.6 were obtained. As will be seen, the mathematical theory of operator semigroups enters into these theories in an intrinsic way.

We may quote Nicolis and Prigogine [1]:

> We thus assert that the *signature of irreversibility* lies in the emergence of a dissipative semigroup description of an appropriately defined markovian process. This is turn leads to one of the deepest questions in physics, namely, can such a markovian process arise from a time-reversible dynamics?

In the mathematical models that are used to explore this view of irreversibility, it is useful to keep in mind the following dichotomies: reversible, irreversible; deterministic, stochastic; conservative, dissipative; group, semigroup; Liouville equation, master equation; hamiltonian, markovian. Roughly speaking, one may regard the first as attributed of a unitary evolution U_t, and the second as attributes of a nonunitary evolution W_t. These evolutions often act on probability distributions (called states) defined over an underlying phase space in which there acts a map S_t taking initial points x into paths $x_t = S_t x$ (called trajectories).

REMARK 1.1.1. The use of dynamical semigroups as a basis for fundamental models of irreversible systems goes back much further, e.g., see Prigogine and Resibois [1], Sudershan, Mathews, and Rao [1], among others. There is a very large literature, which we do not attempt to account for here. The ideas apparently originated from a notion of master equation due to Pauli in the 1920's. An over-simplified statement of the general situation could be the following: given a unitary group evolution U_t generated by the exponentiation of a physically important selfadjoint Hamiltonian operator H, perform a projection (coarse-graining) or a similarity transformation (change of representation) or some other reduction of the system (e.g., Friedrich's Model) in order to study a subsystem of the total physical system of interest. The projected, transformed, or otherwise reduced evolution W_t may or may not be a group (e.g., reversible). For interest in modelling irreversibility, one prefers that it not be. On the other hand, it is very useful that W_t remain a semigroup, and numerous assumptions or approximations have been employed to guarantee that property.

The situation to which we will restrict attention is the following. Let S_t be a semigroup dynamics of point transformation (trajectories) $x_t = S_t x_0$ in a model phase space Ω. For convenience Ω is usually taken compact and with measure $\mu(\Omega) = 1$.

LEMMA 1.1.2 (Koopman). Given a semigroup S_t of point transformations in a topological phase space Ω and an invariant measure μ representing an equilibrium solution, consider the Hilbert space $\mathcal{L}^2(\Omega, \mathcal{B}, \mu)$ and the operators

$$(1.1.1) \qquad V_t f(\omega) = f(S_t \omega)$$

defined for all f in $\mathcal{L}^2$. Then the V_t are isometries, and they are unitary when the S_t are automorphisms.

The V_t are commonly called the Koopman operators induced by the dynamical system S_t. Most early interest centered on the automorphic (measure preserving) case. In that case note the properties carried by V_t in Lemma 1.1.2,

$$(1.1.2) \qquad V_t(1) = 1, \quad V_t \text{ unitary}, \quad V_t f \geq 0 \text{ for } f \geq 0,$$

where 1 denotes the equilibrium solution.

This construction by Koopman [1] ushered in the use of operator theory to study the ergodic properties of dynamical systems. The quickly following important paper of Koopman and Von Neumann [1] observed that for systems with continuous spectra, the states of motion "spread out into an amorphous everywhere dense *chaos*." Later this operator-theoretic approach to dynamical systems was extended to Markov processes, e.g., see Yosida and Kakutani [1].

Within this setting, we may consider Markov semigroups W_t, $t \geqq 0$, on $\mathcal{L}^2(\Omega, \mathcal{B}, \mu)$: one parameter strongly continuous families of operators possessing the properties

$$(1.1.3) \qquad W_t(1) = 1, \quad W_t \text{ a contraction semigroup}, \quad W_t f \geq 0 \text{ for } f \geq 0.$$

Associated to the semigroup W_t is the transition probability $P_t(\omega, \Delta) = W_t \chi_\Delta(\omega)$ from the point ω into the measurable set Δ in time t. Conversely, such one parameter probabilities $P_t(\omega, \Delta)$ which also satisfy the Chapman–Kolmogorov condition $P_{t+s}(\omega, \Delta) = \int_{\Omega} P_t(\omega, d\omega') P_s(\omega', \Delta)$ induce a Markov semigroup $W_t f(\omega) = \int_{\Omega} P_t(\omega, d\omega') f(\omega')$.

In the following, given a unitary group evolution U_t, two transformed evolutions are considered, namely,

$$(1.1.4) \qquad\qquad W_t = PU_t P$$

where P is a projection satisfying

$$(1.1.5) \qquad\qquad P(1) = 1, \quad P \text{ is orthogonal}, \quad Pf \geq 0 \text{ for } f \geq 0,$$

and

$$(1.1.6) \qquad\qquad W_t = \Lambda U_t \Lambda^{-1}$$

where Λ is a bounded invertible similarity transformation satisfying

$$(1.1.7) \qquad\qquad \Lambda(1) = 1, \quad \Lambda \text{ preserves normalization}, \quad \Lambda f \geq 0 \text{ for } f \geq 0.$$

Recall a projection P is orthogonal in a Hilbert space iff $P = P^*$. That Λ preserves (probability) normalization means $\int_{\Omega} \Lambda f d\mu = \int_{\Omega} f d\mu$ for all $f \geq 0$, equivalently $\Lambda^*(1) = 1$.

LEMMA 1.1.3. Projected evolutions (1.1.4) are semigroups iff P projects onto a nonregenerating subspace, i.e.,

$$(1.1.8) \qquad\qquad PU_s P^{\perp} U_t P = 0, \quad \text{all } t \geq 0, \ s \geq 0.$$

Transformed evolutions (1.1.6) are always semigroups.

Proof. Straightforward. See Gustafson [2].

If P commutes with U_t, then W_t remains a group, so the P to be appropriate are of the coarse graining type and are not to commute with U_t, so that determinism will be lost. In like vein, we will see below that if Λ^{-1} is also positivity preserving, then W_t remains deterministic, so such Λ are not appropriate for the irreversibility models.

REMARK 1.1.4. The notion of nonregenerative subspace came from a model for decaying subspaces, see Gustafson and Misra [1]. There it was shown that such nonregenerative subspaces cannot be used to model unstable quantum particles. For more implications related to Lemma 1.1.3, see Gustafson [2].

Now we may state results of the Brussels school mentioned above. These give sufficient conditions for a change from deterministic dynamics to a probabilistic description.

THEOREM 1.1.5. Given a deterministic (measure preserving) dynamics S_t in a phase space Ω described by its Koopman unitary evolution U_t on the state space $\mathcal{L}^2(\Omega, \mathcal{B}, \mu)$, if the dynamical system S_t is a Kolmogorov dynamical system then the projected coarse-grained evolution $W_t = PU_tP$ is an irreversible Markov semigroup. The assumptions on P are those of (1.1.5) and in addition the two properties

$$(1.1.9) \qquad PU_tf = PU_tf' \text{ for all } t \Rightarrow f = f'$$

$$(1.1.10) \qquad PU_t = W_t^*P \text{ for } t \geq 0.$$

The irreversible semigroup W_t satisfies (1.1.3) and in addition the two properties

$$(1.1.11) \qquad W_t^*(1) = 1$$

$$(1.1.12) \qquad \lim_{t \to \infty} \|W_t^*f - 1\| = 0 \text{ monotonically}, \; \forall \, f \geq 0, \; \int_\Omega f d\mu = 1.$$

Proof. See Misra and Prigogine [2]. Preliminary versions appeared in Misra and Prigogine [1], Misra, Prigogine and Courbage [1], [2]. The earlier papers worked with specific simpler underlying dynamics such as iterations S^n of the baker's transformation

$$(1.1.13) \qquad S(x,y) = \begin{cases} (2x, y/2), & 0 \leq x < 1/2 \\ (2x - 1, (y+1)/2), & 1/2 \leqq x \leq 1 \end{cases}$$

on the unit square $[0, 1] \times [0, 1]$. The theory was then extended to Bernoulli systems and finally to Kolmogorov systems.

Kolmogorov systems (e.g., see Cornfeld, Fomin, and Sinai [1] or the book Lasota and Mackey [1]) are higher unstable invertible dynamical systems S_t, t real or integer, characterized by a measurable partition ξ, the K-partition, of the phase space Ω, which evolves asymmetrically, with the following properties:

(1) ξ is progressively refined: $S_t\xi < S_{t'}\xi$, $t < t'$

(2) ξ approaches the finest point partition ν in the far

(1.1.14)   future: $\displaystyle\bigvee_{-\infty < t < \infty} S_t\xi = \nu$

(3) ξ approaches the coarsest one cell partition ε

in the far past: $\displaystyle\bigwedge_{-\infty < t < \infty} S_t\xi = \varepsilon.$

Examples of Kolmogorov systems include the infinite ideal gas, hard rods systems, motion of a billiard ball on a table with a convex obstacle, hard sphere gases, geodesic flow on a manifold of negative curvature. The K-partition ξ generalizes the stable manifold of hyperbolic dynamical systems. For example, for the baker's transformation (1.1.13), the cells of the K-partition are the vertical lines. The trajectories emanating from points on such a line become more and more indistinguishable as the iterations S^n continue. Meanwhile points on horizontal lines (the unstable manifold) diverge exponentially.

Generally we will use the terms K-flow and K-systems somewhat indiscriminately for both discrete and continuous time Kolmogorov systems. Generally K-systems connote discrete time and K-flows continuous time parameters. These structures should not be confused with a Kolmogorov flow occurring in the theory of fluid dynamics.

Theorem 1.1.5 addresses the process of coarse graining P and the resulting semigroup (1.1.4). The other Λ transformation of a deterministic group U_t to a stochastic semigroup W_t (1.1.6) was obtained at about the same time (although their historical antecedents were quite different, see Remark 1.1.7 below). In the early basic paper Misra, Prigogine, Courbage [1], it was proposed that a unitary evolution U_t induced from an underlying dynamics S_t be considered "equivalent" to a stochastic evolution W_t provided that both leave invariant the same "physical" measure μ on the phase space Ω and provided that there exist a "change of representation" implemented by a bounded operator Λ on $\mathcal{L}^2(\Omega, \mathcal{B}, \mu)$ satisfying certain appropriate conditions.

THEOREM 1.1.6. Given a deterministic (measure preserving) dynamics S_t in a phase space Ω and the induced Koopman evolution operators $U_t f(\omega) = f(S_{-t}\omega)$ on density functions f in $\mathcal{L}^2(\Omega, \mathcal{B}, \mu)$, if S_t is a Kolmogorov dynamical system and there exists a suitable change of representation operator Λ, then the transformed evolution $W_t = \Lambda U_t \Lambda^{-1}$ is an irreversible Markov semigroup. The assumptions on Λ are those of (1.1.7) and in addition the two properties

$$(1.1.15) \qquad\qquad \Lambda^{-1} \quad \text{is densely defined}$$

$$(1.1.16) \qquad\qquad \Lambda U_t = W_t^* \Lambda \quad \text{for} \quad t \geq 0.$$

The irreversible semigroup W_t satisfies (1.1.3) and in addition the two properties (1.1.11) and (1.1.12).

Proof. See Goldstein, Misra, and Courbage [1]. The earlier versions did not treat the full question of all K-flows.

REMARK 1.1.7. The coarse-graining P is viewed not as an objective physical phenomena but rather as a representation of the fact that we are always ignorant of the exact dynamical state of a system. In other words, any act of measurement implements at best a conditional expectation, based on finite limits of precision. As is well known, the measurement theory for quantum mechanics exhibits many exasperating examples of this viewpoint (for a spectrum of discussions, see Ludwig [1], M. Mackey [1], and the literature cited therein). The change of representation Λ above is viewed as a symmetry breaking (of time) from which one obtains an exact Markovian master equation (in forward time). Both the coarse graining and the representation change do not depend upon earlier theories which assumed special approximation schemes, such as those employing nonphysical weak coupling limits. The key idea in the approach of the Brussels school is to recognize and use the intrinsic randomness of the underlying highly unstable dynamical system S_t and the ways in which it may induce a symmetry breaking and resulting probabilistic description.

REMARK 1.1.8. Coarse graining is a concept both valued and vilified in contemporary scientific literature. See for example its use by Penrose [1] and its criticisms in Coveney and Highfield [1]. Its difficulty is its subjective nature, coupling the experiment with the observer. However its use here is quite objective, for example the P of Theorem 1.1.5 is usually taken to be the conditional expectation over the cells of the K-partition ξ. For the K-flows that appear in our models, one usually has essentially no choice in how to choose P. To the author's mind, the most limiting aspect of the coarse graining transformation from deterministic to stochastic description lies hidden in Lemma 1.1.3: the fact that there can be no further interaction of the decay products in the orthogonal subspace, with the reduced evolution. For this reason, it was proposed in Gustafson [1],

[2] that the semigroup property in W_t should be given up, or at least relaxed for example to some wider stochastic concept (e.g., martingalian, which in spirit is quite close to K-flow), in order to better fit the physics or better fit the actual loss of information during the physical evolution of the process, or to other stochastic structures which permit some memory effects.

REMARK 1.1.9. The transformation Λ appeared first, Misra [1], as a square root of a Lyapunov variable M. Later, Misra, Prigogine, Courbage [1], Λ was taken to be a function of T, $\Lambda = h(T)$, where T is an operator of "internal time" constructed as an operator dual (in the sense of canonical commutation relations) to the selfadjoint infinitesimal generator H of the unitary evolution U_t. The essentials of this construction go back to Gustafson and Misra [1].

1.2. Three Converses. We will present here, in the form of three converses, some results which clarify some aspects of the irreversibility theory described in the previous section. A main conclusion: that theory is essentially limited to underlying dynamics S_t which are Kolmogorov flows.

The first converse arose out of a conjecture in Misra, Prigogine, Courbage [1] that the transformed evolution $W_t = \Lambda U_t \Lambda^{-1}$ should not be positivity preserving for both positive and negative t. Because both Λ and U_t are positivity preserving, this conjecture focuses attention on Λ^{-1}. That Λ^{-1} cannot be positivity preserving follows easily once the following converse to Koopman's Lemma 1.1.2 is established.

LEMMA 1.2.1 (Koopman Converse). Let V_t be a strongly continuous one parameter group of operators on $\mathcal{L}^2(\Omega, \mathcal{B}, \mu)$ such that for all $-\infty < t < \infty$ the conditions (1.1.2) are satisfied. Then there are underlying measure preserving point transformations S_t such that

$$(1.2.1) \qquad (V_t f)(\omega) = f(S_t \omega)$$

and such that S_t is a weakly measurable group.

Proof. See Goodrich, Gustafson, Misra [1] for more details. Here we will sketch the essentials of the proof. By the parallelogram law for a Hilbert space, if V is an isometry and E_1 and E_2 are disjoint sets in the phase space Ω, then for their characteristic functions $\chi(E_1)$ and $\chi(E_2)$ we have

$$
\begin{aligned}
\|V[\chi(E_1) + \chi(E_2)]\|^2 &= \|\chi(E_1)\|^2 + \|\chi(E_2)\|^2 \\
&= \|V\chi(E_1)\|^2 + \|V\chi(E_2)\|^2
\end{aligned}
$$
$(1.2.2)$

Hence

$$(1.2.3) \qquad 2\mathrm{Re}\,\langle V\chi(E_1)),V(\chi(E_2))\rangle = 2\mathrm{Re}\int_\Omega V(\chi(E_1)(\omega)V(\chi(E_2)(\omega))d\omega = 0.$$

But because V is positivity preserving, the integrand is nonnegative, and hence zero a.e. Therefore $V(\chi(E_1))$ and $V(\chi(E_2))$ have disjoint (up to a null set) supports. From this a standard σ-homomorphism construction may be employed to arrive at an underlying point transformation S for which (1.2.1) holds. The unitarity of V guarantees that S is measure preserving. This construction applies for all t in Lemma 1.2.1. That S_t is weakly measurable is well known to be equivalent to the strong continuity of the family V_t.

COROLLARY 1.2.2. A closed densely defined operator Λ cannot satisfy conditions (1.1.7) and

$$(1.2.4) \qquad \mathcal{D}(\Lambda) \quad \text{contains all characteristic functions of Borel sets}$$

$$(1.2.5) \qquad \mathcal{R}(\Lambda) \quad \text{contains all characteristic functions of Borel sets}$$

$$(1.2.6) \qquad \Lambda^{-1}f \geq 0 \quad \text{for} \quad f \geq 0$$

unless Λ is induced from a measure preserving point transformation of the phase space Ω.

Proof. Similar to that of Lemma 1.2.1. As there, one uses the fact that the characteristic functions are dense in the state space.

Corollary 1.2.2 answers in the affirmative the conjecture of Misra, Prigogine, and Courbage [1]. More generally it affirms the principle, often used intuitively in chemistry, that a forward moving (e.g., chemical) process that loses information cannot be reversed, see Gustafson, Goodrich, and Misra [1]. In terms of the Brussel's School theory, it says that the reverse transformation $f \to \Lambda^{-1}f$ which would lead from the reduced description $W_t = \Lambda U_t \Lambda^{-1}$ back to the deterministic description U_t, must violate probability preservation in either forward or backward time.

REMARK 1.2.3. One will notice some discrepancy in the literature between the occurrence of W_t or instead W_t^* to represent the reduced evolutions PU_tP or $\Lambda U_t\Lambda^{-1}$. Although some care is needed, here for simplicity we have just used W_t in all cases. When W_t^* is employed, it should be thought of as the adjoint of a Markov family M_t. The choice of W_t or W_t^* depends on the convention of S_t or S_t^{-1} in the underlying dynamics.

REMARK 1.2.4. The converse Lemma 1.2.1 to Koopman's Lemma 1.1.2 could be obtained in other ways. See the discussion in Gustafson and Goodrich [1]. To summarize, we were led to

Lemma 1.2.1 from a positivity preserving, physical, $\mathcal{L}^2$ point of view. One could instead have used a multiplication preserving, square root $\mathcal{L}^1 \cap \mathcal{L}^2$ point of view. Or one could have used a multiplication operator, Von Neumann algebra, $\mathcal{L}^\infty$ point of view.

REMARK 1.2.5. Lemma 1.2.1 can also be viewed as a theorem of Banach–Lamperti type. Banach [1] proved that for $\mathcal{L}^p$, $1 \le p < \infty$, $p \ne 2$, all isometries $\|Vf\| = \|f\|$ come from underlying point transformations. Lamperti [1] extended this to $0 < p < 1$. A Radon–Nikodym derivative h may enter,

$$(1.2.7) \qquad\qquad (Vf)(\omega) = h(\omega)f(S\omega)$$

but that is a technical point only. Both S and h are determined uniquely. The Banach–Lamperti theorem is generally false for $p = 2$. The merit of Lemma 1.2.1 is that it singles out the positivity preserving isometries as the interesting (e.g., physical) ones in Hilbert space, in which there are many isometries.

The second converse, Goodrich, Gustafson, Misra [2], proves that the Prigogine, Misra, Courbage et al. theory is essentially limited to an original deterministic dynamics S_t which is of Kolmogorov flow type. In other words, the prior developments of the theory, in which S_t was presumed to be first baker, then Bernoulli, then more generally Kolmogorov, are in fact necessarily limited to the latter dynamical systems.

THEOREM 1.2.6 (Kolmogorov Converse). Let U_t be a reversible unitary evolution on $\mathcal{L}^2(\Omega, \mathcal{B}, \mu)$ induced by a measure preserving dynamics S_t in Ω. For a projected coarse-grained evolution $W_t = PU_tP$ to be an irreversible Markov semigroup, P satisfying the conditions (1.1.5), (1.1.9), (1.1.10) and W satisfying the conditions (1.1.3), (1.1.11), (1.1.12), it is necessary that S_t be a Kolmogorov flow.

Proof. See Goodrich, Gustafson, Misra [2] for more details. Here we sketch the essential steps in the proof. The key to the proof is the canonical commutation relations, which we state below as Lemma 1.2.7. For the proof of Theorem 1.2.6 we wish to establish a system of imprimitivity related to the unitary evolution U_t. Let $P_{-t} = U_{-t}PU_t$. Then $P_0 \equiv P$ and all P_t are conditional expectations with ranges $\mathcal{R}(P_t)$ in $\mathcal{L}^2(\Omega, \mathcal{B}_t, \mu)$, where the $\mathcal{B}_t$ are σ-subalgebras of $\mathcal{B}$. It may be checked that $\mathcal{B}_t = S_t(\mathcal{B})$. Having identified the P_t, we may now proceed to establish the system of imprimitivity

(1.2.8) $$U_t^* P_\lambda U_t = P_{\lambda - t}$$

as follows. From $P_\lambda = U_\lambda P U_{-\lambda}$ we have $P_\lambda U_t = U_\lambda P U_{-\lambda} U_t = U_\lambda P U_{t-\lambda}$, so $U_t^* P_\lambda U_t = U_{\lambda-t} P U_{t-\lambda} = P_{\lambda-t}$.

The properties (1), (2), (3) of (1.1.14) of a Kolmogorov flow follow. The property (1) of progressive refinement follows from the intertwining condition (1.1.10), which orders the system. The property (2) of completeness follows from the resolution condition (1.1.9). The property (3) of emptiness of the infinite remote past follows from the approach to equilibrium condition (1.1.12).

The imprimitivity relation (1.2.8) was useful in the above proof. For later reference we will need more explicit aspects of the canonical commutation relations, which we now state as a lemma. To prepare the lemma, let

(1.2.9) $$p = -i\frac{d}{dx} = \quad \text{momentum operator in} \quad \mathcal{L}^2(\mathbb{R}^1)$$

(1.2.10) $$q = x = \quad \text{position operator in} \quad \mathcal{L}^2(\mathbb{R}^1)$$

Any pair H_1, H_2 of selfadjoint operators unitarily equivalent to p and q are called a Schrödinger couple.

LEMMA 1.2.7 (Stone-Von Neumann). Each of the following conditions is sufficient for H_1, H_2 to be a Schrödinger couple.

(1.2.11) $\quad U_t = e^{itH_1}, \; V_s = e^{isH_2}, \; U_t V_s = e^{its} V_s U_t$ [Weyl form]

(1.2.12) $\quad U_t = e^{itH_1}, \; H_2 = \int \lambda dE_\lambda, \; U_t^* E_\lambda U_t = E_{\lambda+t}$ [Schrödinger form]

(1.2.13) $\quad [H_1, H_2] = -iI$ plus domain conditions [Heisenberg form]

Proof. See Putnam [1].

REMARK 1.2.8. This lemma has a long and rich history, with far-reaching implications and connections to many parts of mathematics and physics. For instance, the Weyl form (1.2.11) may be regarded as a truncation of the Baker–Campbell–Hausdorff theorem for the exponentiation of operator sums. The Schrödinger form (1.2.12) is a version of imprimitivity systems from the theory of group representations. The Heisenberg form (1.2.13) is the uncertainty relation from quantum mechanics. An example of the needed domain conditions (Rellich–Dixmier): $H_1^2 + H_2^2$ be essentially selfadjoint.

The third converse, Antoniou and Gustafson [1], asks when an irreversible Markov semigroup W_t of the Brussels School type may be "lifted" to a unitary reversible superdynamics U_t induced by an underlying measure preserving point transformation. In other words, the Prigogine et al. theory showed that a Kolmogorov deterministic dynamical system U_t may be projected or transformed to a probabilistic description W_t. We now ask when a probabilistic description W_t may be converted to a deterministic dynamics.

THEOREM 1.2.9 (Prigogine Converse). The Misra–Prigogine–Courbage coarse-grained probabilistic semigroup $W_t = PU_tP$ possesses a natural positive dilation to a deterministic Kolmogorov dynamical system.

Proof. See Antoniou and Gustafson [1] for more details. Here we sketch the main lines of the proof, which relies heavily on dilation theory. In particular, it may be checked that the Prigogine et al. coarse grained semigroups $W_t = PU_tP$ possess six properties (specifically, (α) through (ζ) below) which we will call those of exact Markov semigroups. It may then be shown, using the dilation theory of exact dynamical systems, Rokhlin [1], that such semigroups may be dilated in a natural way to a positive Koopman group evolution on an extended measure space Ω_e.

REMARK 1.2.10. We may say that there are three dilation theories. The first and most well known is due to Halmos, Sz. Nagy, Foias, Schaeffer, Naimark, and others. See the account in Gustafson and Rao [1]. This theory dilates the Hilbert space, and then the operators thereupon. In physical terms we may say that it dilates the probabilities. The second dilation theory is that of Kolmogorov–Rokhlin, as mentioned above. That theory dilates the point dynamics S_t. The third dilation theory, that of Akcoglu, Sucheston, and others, dilates the measures. We would like to note that, upon closer inspection, parts of the Sz. Nagy theory may also be seen to depend upon the extension of a measure. Use of the more familiar Sz. Nagy–Foias et al. dilation theory also provides a unitary dilation of any contractive semigroup, see Sz. Nagy and Foias [1], Halmos [1], Davies [2]. However, to show the unitary dilation V_t corresponds to a Kolmogorov system, we need to show that V_t is positivity preserving and that the projection P onto the original Hilbert space is also positivity preserving. Although this can be shown for some W_t, it does not seem to be true in general. Moreover the componentwise positivity resulting from a Sz. Nagy–Foias dilation construction is not naturally connected to the positivity of the particular Markov semigroup W_t. Attempts to apply the Sz. Nagy dilation construction to Markov processes associated with background Brownian motion and thus "make a heat bath," see Lewis and Thomas [1], Davies [1], do not clarify the positivity question either. They just eliminate instead the concept of positivity from Kolmogorov systems.

REMARK 1.2.11. Even though Theorem 1.2.9 obtains a Kolmogorov unitary dilation family V_t from W_t, the actual construction of these dilations is a computationally difficult task, because we would need to know all inverse paths of the extended trajectories. Due to the chaotic nature of exact extensions, we see that we cannot recover the loss of information caused by the original coarse-graining. However, compare our discussion in the next section.

One may extend the above converses to more general Markov semigroups. The results presented here will appear in Antoniou and Gustafson [4] and Antoniou, Gustafson and Suchanecki [1]. The question posed is: given an arbitrary Markov semigroup M_t with some of the properties mentioned above, but with no idea as to any unitary dynamics V_t from which it may (or may not) have arisen through some contraction of description, when can M_t be dilated to a deterministic evolution on a larger space?

It is useful to organize certain of the properties of Markov semigroups which we have encountered above. A stationary Markov process on the phase space $(\Omega, \mathcal{B}, \mu)$ is described by transition probabilities $Q_t[\Delta \mid \omega]$ from the point ω in Ω into the measurable set Δ in $\mathcal{B}$ in time t, $t \geq 0$. The probability densities $f(\omega)$ evolve according to the Markov semigroup M_t on $\mathcal{L}^2(\Omega, \mathcal{B}, \mu)$ defined by

$$(1.2.14) \qquad \int_\Delta d\mu(\omega) M_t f(\omega) = \int_\Omega d\mu(\omega) f(\omega) Q_t[\Delta \mid \omega]$$

for any measurable set Δ in $\mathcal{B}$. We consider the following properties M_t may possess.

(α) M_t are contractions on $\mathcal{L}^2(\Omega, \mathcal{B}, \mu)$

$$\|M_t f\|^2 \leqq \|f\|^2$$

(β) M_t preserve probability densities

$$M_t f \geq 0 \quad \text{if} \quad f \geq 0$$

(γ) M_t preserve the probability normalization

$$\int_\Omega d\mu M_t f = \int d\mu f$$

for all $f \geq 0$. Equivalently: $M_t^*(1) = 1$.

(δ) M_t preserves the equilibrium

$$M_t(1) = 1$$

(ε) M_t irreversibly approaches the equilibrium

$$\|M_t f - 1\|^2 \to 0 \quad \text{as} \quad t \to \infty$$

REMARK 1.2.12. An operator satisfying (β), (γ), (δ) is sometimes called doubly stochastic. Those three conditions imply (α). We also note that (α) and (γ) imply (δ). Here we shall use the term *measure preserving Markov semigroups* when (α) through (δ) are satisfied. When only (α) through (γ) are assumed, we use the term *Markov semigroup*. When (α) through (ε) are satisfied, we shall use the term *irreversible Markov semigroup*. When additionally M_t is the Frobenius–Perron semigroup of exact dynamical systems, we shall use the term *exact Markov semigroup*. This additional property may be stated as follows:

(ζ) M_t are partial isometries dual to the stochastic semigroups W_t on $\mathcal{L}^2(\Omega, \mathcal{B}, \mu)$ defined by

$$W_t f(\omega) = \int_\Omega Q_t[d\omega' \mid \omega] f(\omega')$$

where

$$Q_t[\Delta \mid \omega] = W_t \chi_\Delta(\omega)$$

and where the W_t are isometries

$$\|W_t f\| = \|f\|$$

for all f in $\mathcal{L}^2(\Omega, \mathcal{B}, \mu)$.

Theorem 1.2.9 established that the Misra, Prigogine, Courbage semigroups are exact Markov semigroups arising from projection of some Kolmogorov dynamical system, and that one may dilate such exact Markov semigroups up to a larger Kolmogorov system. One may similarly dilate all Markov semigroups. Although the proofs are too technical to give here, the following results have been obtained.

THEOREM 1.2.13 Every Markov semigroup M_t arises as a projection of a dynamical system in a larger space. If M_t is a measure preserving Markov semigroup, the dilated evolution V_t is a Kolmogorov system.

Proof. See Antoniou and Gustafson [4], Antoniou, Gustafson and Suchanecki [1]. For discrete Markov semigroups M_n, a dilation theory due to Akcoglu [1] and Akcoglu and Sucheston [1] was employed. For continuous Markov semigroups M_t, we also needed an extension construction for cylindrical algebras due to Ionescu–Tulcea [1]. In both cases we employed also elements of the Kolmogorov–Rokhlin dilation theory.

REMARK 1.2.14. It would be desirable to have such dilation constructions which somehow more adequately take account of the specific physics taking place. Extension of measures although ingrained in the foundations of probability theory, does not do this. The Kolmogorov–Rokhlin construction best retains the dynamics, hence the physics, but our general impression is still one of some dissatisfaction with present dilation theories for specific application to physical dynamics.

REMARK 1.2.15. For M_t a measure preserving Markov semigroup, the dilated evolution V_t when projected back to the semigroup M_t is an averaging over the generating partition ξ of the Kolmogorov system, and not over the partition ξ_0 considered in the Misra, Prigogine, Courbage constructions. Thus only exact Markov semigroups arise as projections from that theory.

REMARK 1.2.16. A specific converse or dilation theory for the Λ change of representation reduced descriptions has not yet been developed. It must be admitted that the present Λ transformation theory does not yet take into account enough specific physics to determine exactly a unique Λ.

1.3. Irreversibility and Second Laws. We would like to conclude this chapter, in which questions of reversibility versus irreversibility entered implicitly throughout, with some discussion, commentary, conjecture about time ordering in nature, and about the degree of appropriateness of mathematical models which attempt to explain such physically motivated ordering. In spite of all the study of this question for over a hundred years by some of the very best minds, the basic question of time ordering remains unresolved.

REMARK 1.3.1. There is a vast literature concerned with these questions. Beyond the famous books Hawking [1] and Penrose [1], we suggest the recent popularization Coveney and Highfield [1] and the literature citations therein. Most attention to time-ordering and irreversibility has come from the standpoint of thermodynamics, where classical experiments early on led to the formulation of the Second Law of Thermodynamics: the law (assumption) that total entropy always increases. This statement presupposes that one is in an irreversible transition. The latter statement presupposes that an ordering of time has been adopted.

REMARK 1.3.2. We have no intention of giving any full discussion of the Second Law. It is stated in many forms and various contexts and in discourses ranging from fundamental physics to pure philosophy to engineering heat transfer applications. Historically and in the context of steam engine technology, the concern was (Second Law) that work converted (wastefully) to heat, but not, unfortunately, vice versa. Feynman [1] puts the (thermodynamic) Second Law succinctly (we paraphrase slightly): the total entropy of the total system increases iff the total change of the system is irreversible. Stated thusly, the emphasis of understanding the Second Law shifts to the meaning of 'irreversible'. One finds fewer conclusive discussions of the latter term. One approach, which have we followed to some extent this chapter, was that if an evolution V_t of states was induced by an underlying set transformation S_t in phase space, and if S_t was not invertible, then V_t was irreversible. However, it need not follow that entropy increase. Moreover, reversible dynamical systems (invariant under the simultaneous variable change $t \to -t$, $\omega \to S\omega$) need not be conservative. Invertible maps can be dissipative (e.g., Henon system). Thus care must be taken concerning relations between the semantic dichotomies mentioned at the beginning of this chapter.

For the purposes of this section, we shall suppose that irreversibility is the foundational second law, that second laws may therefore be of different forms in different contexts, and that physical or informational or other manifestations of irreversibility, such as entropy, are secondary and context-dependent. Our main goal in this section is to advance a proposed new view of physical irreversibility.

The quotation from Nicolis and Prigogine [1] at the beginning of this chapter can be rephrased in operator-theoretic terms as: does a Markovian description in and of itself imply physical time-irreversibility? In Gustafson [7] the following principles were formulated, within a general framework of operator spectral states. Examples will be given below.

PRINCIPLE OF CHOICE: Time is a binary choice. This may come into implementation by semigroup choice, spectral monotonicity choice, continuation sheet choice, eigenfunction space choice, shift direction choice, adjoint or conjugate choice, phase space-state space choice, upper or lower rigging choice, preference for regularity choice, preference for probability choice.

PRINCIPLE OF (MACROSCOPIC) REGULARIZATION: Nature generally prefers regularization.

Entropy increase sometimes coincides with this regularity preference.

PRINCIPLE OF PROBABILITY PREFERENCE: Nature generally prefers to preserve probability densities. When density amplitude exceeds one, information is lost. This may signal an irrevocable irreversibility at a fundamental level.

To these three principles, we here add a fourth. In some sense, this is the most general. It is the 'other half' of the Principle of Regularization.

PRINCIPLE OF (MACROSCOPIC) REFINEMENT OF DETAIL: Detail refinement is both a manifestation and a cause of irreversibility. Elaboration of detail in the finer scales accompanied by regularization of detail in the coarser scales is an extensive second law.

Examples of the Principle of Choice fill the literature of mathematical physics. Whenever a group evolution becomes that of a semigroup, the Principle of Choice has been exercised, either advertently or inadvertently. If you put all the poles in the upper half plane, you have invoked it. If you use Hardy space theory, or inner-outer functions, you have probably invoked it. If you go to the theory of distributions or Gelfand triples, your choice of ϕ below ϕ' has ordered time. See Gustafson [7] for further discussion.

It is important to make clear that the principle that time (direction) is a binary choice, is a statement about models. For example, if you (or some occurrence in nature) chooses the upper half plane over the lower half plane, that is a binary event. One can say that the symmetry between past and future is broken. But only in the models.

Let us bring this point home by recalling the usual experiment performed in a first statistics course. Steel balls are dropped into a grid of nails protruding from a backing. The balls form vertical column histograms approaching a final bell-shaped distribution. The individual trajectories have contributed to a statistical description. If you remember only the evolving distributions for example, only its average location and dispersion, i.e., only the first and second moments, of course you cannot recover the individual trajectory information. But that is not irreversibility. Given a large enough computer, you can retain all individual trajectory information. You can even do this if you allow balls to collide due to "batch processing." Then you add bit tags to keep a record of such events. The choice in going to a statistical description is that of going from a large computer memory to a small one. Given a large enough computer, you can even run the trajectories backward to their initial positions.

Thus we do not agree that irreversibility is an emergent property of processes in which we are forced to go to ensemble rather than particle description. If one loses information, it is because you do not keep track of all evolving relations among the individual trajectories. This applies also to maps S for which S^{-1} is not single valued. The fact that S^{-1} is a relation and not a function is not irreversibility. The often proposed "holistic" dogma that the whole is more than the sum of its parts is also sometimes misconstrued. The whole is exactly its parts plus all relations among them.

It must be admitted that sometimes we do not know all of those relations. This is the crux of the matter. For example postulating a stochastic background for a physical theory, moreover that it be Markovian, is tempting and convenient. Then the view that the physical world and its transitions are always subject to tiny inherent fluctuations replaces the notion of irreversibility with that of unpredictability. Propositions of irreversibility are now transferred to a finer level: what are the mechanisms and nature of such inherent fluctuations or other forms of randomness?

Thus in our view, increasing disorder usually associated with increasing entropy is really increasing order coupled with a refusal or inability to keep track of it. The principle of macroscopic regularization is in fact a duality with the principle of microscopic refinement.

A simple model of this duality is the progression of the Haar basis from the indicator function 1 on the unit interval to ± 1 indicator functions on successive diadic partitions of the unit interval to extremely finely oscillating ± 1 indicator functions which in their blurred refinement limit resemble more and more the indicator function 1 on the whole interval (along with its mirror image -1 below). Great microscopic refinement has taken place while approaching a macroscopically regular final equilibrium.

REMARK 1.3.3. For more explanation of the Principle of Regularization, see Gustafson [7]. For example, as emphasized in the book Gustafson [3], in the classical Dirichlet problem on the unit disk it is in fact a desire for interior C^1 regularity in the solution $u(r,\theta)$, and not the conventionally stated periodicity requirement, that determines a solution. The periodicity gives the (Humankind) convenient and specific Fourier trigonometric expansions. The regularity (Nature) is more basic, and corresponds to just stretching a smooth minimal stress most regular surface across the boundary. A less elementary example of the Principle of Regularization is found in the equations of gas dynamics. How to uniquely continue the solution 'downstream' introduces the notions of Rankine–Hugoniot jump conditions. The latter are often justified in terms of entropy increase at a shock. As argued in Gustafson [3], [7], this may be seen as an instance of the Principle of Regularization. The Rankine–Hugoniot conditions select the most regular weak solution. The situation is seen more explicitly in the Riemann shock tube problem: one treats existence and uniqueness considerations by specifying that shocks will go into the forward space-time cone and rarefaction fans (more regular) will go into the backward space-time cone.

REMARK 1.3.4. The Principle of Refinement of Detail is actually one of duality with regularization. Regularization is, macroscopically, a grinding down process, or a mixing process, a transfer from organized form to disorganized form, e.g., from crystalline structure into heat through friction. But microscopically, detail is increasing. The heat is like Brownian motion of a very detailed kind. It may be stochastic but it is refinement of detail. Similarly, in fluids, in order to fill all possible space within a region of flow, a refinement of detail must take place, see Gustafson [4]. Macroscopically, the fluid seeks a nonequilibrium but sustainable state. Such macroscopical regularization requires concomitant microscopic fine structures.

The Principle of Probability Preference is an outgrowth of a proposed fundamental new understanding of quantum scattering events in the continuous spectrum, Gustafson [7]. In this new theory, when a spectral function $\langle E(\lambda)\phi, \phi \rangle$ at a continuum energy E_0 absorbs an incoming photon, the gain (slope of $\langle E(\lambda)\phi, \phi \rangle$) may be increased so that the derivative of $\langle E(\lambda)\phi, \phi \rangle$ exceeds one. This is a violation of probability density and triggers a spontaneous emission. The Principle of Probability Preference may be seen as an instance of the duality of the Principles of Regularity and Refinement, even if we are unable to know the detailed mechanisms of the latter in every situation.

Because Second Laws are not understood at the level of quantum mechanics, let us consider a steepening spectral family $\langle E(\lambda)\phi, \phi \rangle$ over a segment of continuous spectrum. This occurs in quantum mechanics as a spectral concentration corresponding to a pole in a scattering cross section. Such a steepening is sometimes called a pseudoeigenvalue in the continuum, corresponding to a Green's function singularity nearby in either the upper or lower half plane, depending on your "time" convention. For purposes of a general discussion, let x denote the continuum parameter (e.g., energy λ, frequency k, etc.) and y denote the probabilistic spectral distribution function (e.g., $\langle E(x)\phi, \phi \rangle$, normalized power spectrum cumulative distribution $\int_0^x |\hat{f}(k)|^2 dk$, etc.) for a given state

ϕ. Let us take as our model spectral family $E(x)$ the important threshold function, the sigmoid

$$(1.3.1) \qquad\qquad y(x) = \frac{1}{1 + e^{-\beta(x-\theta)}}$$

where θ is called the threshold and β is called the gain. There are multicultural motivations for this important function (see, e.g., Chapter 3) but a good one is that of transconductance amplifier: you input a voltage x relative to a bias θ and enjoy an enhanced current output y. For simplicity we may take θ to be zero, and arrive at the important equation characterizing the sigmoid,

$$(1.3.2) \qquad\qquad y'(x) = \beta y(x)(1-y)(x).$$

The sigmoid $y(x)$ is an absolutely continuous probability distribution and its derivative $y(x)$ is a corresponding probability density function *provided that* its gain does not exceed 4.

Before continuing with our sigmoid spectral distribution model, let us distinguish the notion of probability density violation from what might at first appear to be equivalent notions in other guises. First, the whole literature of operator theoretic statistical mechanics posits that the positivity property $\rho(x) \geqq 0$ of state functions defined on a phase space should be preserved under all operations on the state space. This is a principle of probability preservation. Should some $\rho(x)$ go negative, then a complementary probability $1 - \rho(x)$ will violate probability density in our sense. But this notion, that of a space of positive probability densities, comes just from the model setup in which the positive functions $\rho(x)$ are *defined* to be the density states. The idea of probability density violation embodied within the Principle of Probability Preservation is different, comes from different motivations, and drives at the question: what "happens" to the "information" in a probability density function that is dynamically driven "above" the value 1?

As a second source of possible overlap with what is known, in the scattering theory literature one finds Van Kampen's causality condition. Nussenzveig [1, p. 82] states this as follows. "Let the incoming wave packet be so normalized as to represent one incident particle for $t \to -\infty$. Then the total probability of finding a particle outside of the scatterer at any given moment cannot exceed unity." This condition is derived from an assumed continuity equation for probability flow which is to be probability conserving in the sense of no emission or absorption of probability. Then Van Kampen's postulate is obtained as the complement of a postulate of positive probability densities, in the same way as for the state space models described above. In Van Kampen's formulation it is a complement in the spatial sense, that a particle must be either inside a sphere of radius r, or outside of it, with combined probability one. In contrast, the principle of probability preference operates at a more fundamental level and is deeper than just asking that probabilities be nonnegative. We are interested in fact in those instances in nature in which this preference *cannot* be maintained, is in fact violated, causing something irreversible to happen. This is the conjecture: that it is interesting to inquire as to what "happens" when the peak of the probability density is driven to a value above 1? Does that change the fundamental dynamics of the dynamical system under description? Does it create an irreversibility in an essential dynamical characteristic of the evolution due to "some probability being lost"?

A rather interesting insight may now be seen through the sigmoid distribution, which was introduced at the beginning as a model quantum probabilistic distribution function. The sigmoid derivative $y'(x) = \beta y(x)(1-y)(x)$ corresponds to the logistic iterative map $y_{n+1} = \mu y_n(1 - y_n)$. (Actually, one must be careful with this continuous map–discrete map analogy, for if one carries out a direct discretization of the former, one does not arrive directly at the latter). We recall that the logistic iterative map has been a staple in the modern study of chaotic dynamical systems, see Devaney [1]. As the parameter μ is increased through small positive values, one progresses through a 'bifurcation' sequence of solutions showing increasing complexity. For $\mu > 4$, the orbit of any y which exceeds the value one tends thereafter to $-\infty$. The set Λ of points y which never escape becomes a repelling hyperbolic Cantor set. On it the logistic map can be represented as a shift. The information on the complement of Λ has all gone off to minus infinity.

Returning now to a spectral concentration in a scattering event, consider the schematic Fig. II-1 which depicts a photoeffect in which a "bound-free" transition is caused by an incoming photon event. The spectral function $\langle E(\lambda)\phi, \phi \rangle$ is assumed to be a sigmoid near a continuum energy E_0, both before and after the photon transition event, but a sigmoid with higher gain after the event. When gain becomes too large, the continuum cannot hold the transferred energy, and for those states ϕ the atom undergoes a probability violation. For the interval of time during which the photoeffect is active, this probability violation produces a local chaos in the manner analogous for example to that known for the logistic map.

Continuing, let us look at the Fourier transform of the sigmoid. Since

(1.3.3)
$$\frac{1}{1+e^{-x}} = \tanh\frac{x}{2} + \frac{1}{2}$$

we know that the sigmoid transforms, up to scale factors, as

(1.3.4)
$$f(\lambda) = \frac{1}{1+e^{-\lambda}} \rightarrow -i\operatorname{csch} t + \delta(0).$$

With gain β and threshold included, the transform becomes

(1.3.5)
$$f(\lambda) = \frac{1}{1+e^{-\beta(\lambda-\theta)}} \rightarrow e^{-i2\pi\theta}\frac{\delta(s)}{2} - ie^{-i2\pi\theta}\left(\frac{\pi}{\beta}\right)\operatorname{csch}\left(\frac{2\pi^2}{\beta}s\right).$$

Thus the effect of the threshold $\theta = E_0$ in the energy continuum is a phase change, and the real part of the transform becomes a potential source in the transform domain. Ignoring the phase change, i.e., setting $\theta = 0$, the derivative of the sigmoid transforms as

(1.3.6)
$$f'(\lambda) = \beta f(\lambda)(1-f(\lambda)) \rightarrow -\left(\frac{2\pi^2}{\beta}s\right)\operatorname{csch}\left(\frac{2\pi^2}{\beta}s\right) + i2\pi s\delta(s)$$

in which now the imaginary part of the transform is in the role of source in the transform domain. The gain β enters into the imaginary part of the transform of the spectral distribution $f(\lambda)$ and into the real part of the transform of the spectral derivative $f'(\lambda)$ and accounts for the slope of the developing singularity of the hyperbolic cosecant along those axes, respectively. The gain β of course also acts as a width parameter in the probability density $f'(\lambda)$. Notice that the sigmoid density behaves very much like the Breit–Wigner formula for the partial cross section

(1.3.7)
$$\sigma_\ell(\lambda) \cong \sin^2\delta_\ell(\lambda) = \frac{(\Gamma/2)^2}{(\lambda-E_0)^2 + (\Gamma/2)^2}$$

corresponding to a resonance of angular momentum ℓ with energy peak at E_0 and a pole nearby at $E_0 - i\Gamma/2$ in the lower half plane. As is well known Schiff [1], the line width Γ may be interpreted as the transition probability per unit time for spontaneous emission. Thus the sigmoid gain, β, may also be so interpreted. When probability preference is violated due to too much gain, the radiation energies have exceeded some fundamental informational limit. The reciprocal $1/\beta$ is the time that the quantum system can stay in its upper energy state. Thus any measurement of that upper energy state cannot take longer than the $1/\beta$ lifetime of that state. According to the uncertainty principle

$\Delta t \cdot \Delta \lambda \geq \hbar$, the measurement of that energy state has at best an accuracy of $\beta \hbar$. That uncertainty translates into an emitted line broadening of magnitude β reflecting the shortened lifetime of the higher energy state caused by its measurement.

It was mentioned above that the principle of probability preference may be seen as another instance of the principle of regularity of the preceding section, although all of the mechanics and subtle implications of that statement may not be clear. In the photoeffect photon absorption described above, the regularity of an eigenvalue is exchanged for the less precise but larger bandwidth pseudoeigenvalue and then when gain exceeds four there is a further broadening into the complex plane. Inasmuch as absorption and emission are dual events it is difficult to ascribe an apriori regularity preference of one over the other. Intuitively, both are regularizing events to avoid the collision of two electron trajectories. The view here is that within macroscopic regularizing dynamics, microscopic probability preference may act to minimize the loss of information caused by too much gain at resonant energy levels. More generally, perhaps probability preference may be viewed as nature's way of spreading probability densities to avoid probability violation.

Chapter 2   Scaling Theories

2.1. Multiresolution Analyses. Wavelet analysis as a revitalization of harmonic analysis has provided new bridges to science and engineering. As with Fourier analysis, the most important field of application thus far has been signal processing and pattern analysis. In physics and mathematics the principal new connections of the wavelet theory have been to the theory of group representations, most notably the Weyl–Heisenberg group, to the theory of reproducing kernel Hilbert spaces, most notably the Calderon–Zygmund singular integral operators, and to approximation theory, most notably that of cardinal splines. In the background has always been the connection to the theory of coherent states. Three excellent recent books Daubechies [1], Chui [2], Meyer [1] report on these developments and connections, and will supply the reader with a wide variety of references. See also the survey by Heil and Walnut [1], and the exposés by Strang [1], [2] and Briggs and Henson [1]. In this chapter, three new (related) connections of wavelet structures to other mathematical structures are given: to (1) Kolmogorov dynamical systems, (2) stochastic processes, and (3) Time Operators.

In this first section we recall the salient properties of wavelets as described through their multiresolution analyses, and then establish the connection to Kolmogorov Systems. The next two sections establish the second and third connections just mentioned.

REMARK 2.1.1. The existence of a fundamental connection between Kolmogorov systems and Haar and wavelet-like structures was noted implicitly in 1985 in Gustafson and Goodrich [2]. Wavelet theory can be said to go back to Gabor [1], who established the concepts of phase space analysis of signal processing, and to Haar [1], who established hierarchical complete orthonormal sets independent from differential equations.

The notion of multiresolution analysis can be traced to the work of Meyer [1] and Mallat [1]. Approximation theorists at about the same time were considering related subdivision schemes, Chui [1]. The earlier Laplacian pyramid algorithm due to Burt and Adelson [1] and subband filtering schemes have properties of a multiresolution analysis. Multiresolution analysis has become the essential framework within which to understand and explore wavelet structures. See the recent survey by Jawerth and Sweldens [1].

A multiresolution analysis (MRA) of square integrable functions over the real line consists of a hierarchy of approximations defined as averages on different scales. The finer the scale the better the approximation. If P_n is the orthoprojection corresponding to the approximation at a resolution 2^n, the ranges $\mathcal{H}_n$, $n = 0, \pm 1, \pm 2, \ldots$ of P_n define (Mallat [1], Daubechies [2]) a sequence of Hilbert subspaces of $\mathcal{L}_{\mathbb{R}}^2$ with the properties:

1) $\mathcal{H}_n \subset \mathcal{H}_{n+1}$

2) $\bigcap_n \mathcal{H}_n = \{0\}$

3) $\bigcup_n \mathcal{H}_n$ is dense in $\mathcal{L}_{\mathbb{R}}^2$

4) A function $f(x)$ is in the space $\mathcal{H}_n$ if and only if the scaled function $f(2x)$ is in the space $\mathcal{H}_{n+1}$.

5) There exists a function ϕ in $\mathcal{H}_0$ such that the set of translated functions $\phi_\alpha(x) = \phi(x - \alpha)$, $\alpha = 0, \pm 1, \pm 2, \ldots$ is an orthonormal basis for the space $\mathcal{H}_0$.

We remark that this last property 5) of a MRA is stated in a number of ways in the literature, but all versions either explicitly or implicitly imply that such a function ϕ has been chosen or can be constructed. For example it is sufficient to require that one can find a function $g(x)$ in $\mathcal{L}_{\mathbb{R}}^2$ such that the translated functions $g(x - \alpha)$ form a Riesz basis of $\mathcal{H}_0$. Then one can construct an orthonormal basis $\phi(x - \alpha)$ for $\mathcal{H}_0$ in standard ways.

Condition 5) means that the function ϕ is a cyclic vector for unitary representation of the discrete translation group $U_\alpha \phi(x) = \phi(x - \alpha)$, $\alpha \in \mathbb{Z}$, on the space $\mathcal{H}_0$. As the spaces $\mathcal{H}_n$ are just scalings of the space $\mathcal{H}_0$ (Condition 4)), each scaled function $\phi(2^n x)$ is also a cyclic vector for the unitary representation of the discrete translation group on the space $\mathcal{H}_n$, $n \in \mathbb{Z}$. The function ϕ gives rise therefore to the orthonormal basis

$$(2.1.1) \qquad \phi_{n\alpha}(x) = 2^{\frac{n}{2}} \phi(2^n x - \alpha), \quad n \in \mathbb{Z}, \ \alpha \in \mathbb{Z}$$

of the space $\mathcal{L}_{\mathbb{R}}^2$.

The function ϕ is therefore also a cyclic vector for the unitary representation on $\mathcal{L}_{\mathbb{R}}^2$ of a discrete subgroup of the affine group of translations and scalings

$$(2.1.2) \qquad U_{\alpha\beta} f(x) = \frac{1}{\sqrt{|\beta|}} \, f(\beta^{-1} x - \alpha) \equiv \frac{1}{\sqrt{|\beta|}} \, f((\alpha, \beta)^{-1} x)$$

$\alpha \in \mathbb{R}$, $\beta \in \mathbb{R} - \{0\}$. The affine group $\mathbb{R} \times \mathbb{R} - \{0\}$ acts on the real line as follows:

$$(2.1.3) \qquad x \mapsto (\alpha, \beta) x = \beta(x + \alpha)$$

i.e., first translation by $\alpha \in \mathbb{R}$ and then scaling by $\beta \neq 0$. The group synthesis is

$$(2.1.4) \qquad (\alpha, \beta)(\alpha'\beta') = (\beta\beta', \alpha' + \beta'^{-1}\alpha)$$

The unit and inverse transformations are

$$(2.1.5) \qquad I = (0, 1)$$

$$(2.1.6) \qquad (\alpha, \beta)^{-1} = (-\beta\alpha, \beta^{-1})$$

The function ϕ satisfies the scaling equation

$$(2.1.7) \qquad \phi(x) = \sum_n C_n \phi(2x - k)$$

and can be constructed as a solution of this equation. Although ϕ is usually called scaling function or generating function we shall suggest and use the term *shape function* because ϕ describes the shape of the approximations. This term also occurs in spline and finite element theories with similar connotations.

The successive multiresolution approximations of a function f are expressed through the projections P_n onto $\mathcal{H}_n$:

$$(2.1.8) \qquad P_n f = \sum_{\alpha \in \mathbb{Z}} \langle \phi_{n\alpha}, f \rangle \phi_{n\alpha}$$

with $\langle \phi, f \rangle = \int_{-\infty}^{+\infty} dx \phi^*(x) f(x)$. The detail or new information between two successive approximations $P_n f$ and $P_{n+1} f$ of the functions f is given by the difference $(P_{n+1} - P_n)f$. The range of the projection $P_{n+1} - P_n$ is the orthocomplement $\mathcal{W}_n$ of the space $\mathcal{H}_n$ in $\mathcal{H}_{n+1}$

$$(2.1.9) \qquad \mathcal{W}_n = \mathcal{H}_{n+1} \ominus \mathcal{H}_n, \quad \mathcal{W}_n \perp \mathcal{H}_n$$

The properties (1–5) of the MRA imply that the spaces $\mathcal{W}_n$ are also the scalings of $\mathcal{W}_0$ and that they are mutually orthogonal spaces generating all the space $\mathcal{L}_{\mathbb{R}}^2$:

$$(2.1.10) \qquad \mathcal{W}_n \perp \mathcal{W}_{n'}, \quad n \neq n'$$

$$(2.1.11) \qquad \mathcal{L}_{\mathbb{R}}^2 = \bigoplus_{n \in \mathbb{Z}} \mathcal{W}_n$$

From the shape function ϕ one can construct in a standard way an orthonormal basis for the space $\mathcal{W}_n$ in the form $\{\psi(x - \alpha), \ \alpha \in \mathbb{Z}\}$. The function $\psi(x)$ is called a *wavelet* function in $\mathcal{L}_{\mathbb{R}}^2$. Since the spaces $\mathcal{W}_n$ are obtained from each other by scalings and are mutually orthogonal, the functions

$$(2.1.12) \qquad \psi_{n\alpha}(x) = 2^{\frac{n}{2}} \psi(2^n x - \alpha), \quad \alpha \in \mathbb{Z}$$

for fixed $n \in \mathbb{Z}$ form an orthonormal basis for $\mathcal{W}_n$. The set $\psi_{n\alpha}$, $n \in \mathbb{Z}$, $\alpha \in \mathbb{Z}$ is therefore an orthonormal basis for the whole space $\mathcal{L}_{\mathbb{R}}^2$. For this reason the subspaces $\mathcal{W}_n$ are called the n-wavelet subspaces and the orthonormal basis $\psi_{n\alpha}$ is called the wavelet basis of $\mathcal{L}_{\mathbb{R}}^2$. We may also express the approximation projections P_n of (2.1.8) in terms of the wavelets $\psi_{n\alpha}$:

$$(2.1.13) \qquad P_n f = \sum_{\alpha \in \mathbb{Z}} \langle \psi_{n\alpha}, f \rangle \psi_{n\alpha}$$

In general wavelets are functions ψ in $\mathcal{L}_{\mathbb{R}}^2$ such that the discrete translations and scalings provide orthonormal bases in $\mathcal{L}_{\mathbb{R}}^2$. One adds moreover various conditions such as regularity, vanishing moments, specified decrease at infinity, compact support. For example, the admissibility condition

$$(2.1.14) \qquad \int_{-\infty}^{+\infty} dk \frac{|\hat{\psi}(k)|^2}{|k|} < +\infty$$

with $\hat{\psi}(k) = \int_{-\infty}^{+\infty} \frac{dx}{\sqrt{2\pi}} e^{-ikx} \psi(x)$ implies for regular wavelets that the mean is zero

$$(2.1.15) \qquad \int_{-\infty}^{+\infty} dx \psi(x) = 0$$

REMARK 2.1.2. In connection with moment condition (2.1.15) let us recall that the role of the wavelet ψ is that of an *oscillation function*, for it introduces the needed oscillations into the given shape function ϕ. This was its role classically for Gabor shape function ϕ the Gaussian e^{-x^2} suitably normalized, for which the corresponding wavelet function $\psi(x) = -d^2(e^{-x^2})/dx^2 = (1 - x^2)e^{-x^2}$ suitably normalized is obtained by differentiation. A second classical example for ϕ is the Haar function $1_{[0,1]}(x)$, for which $\psi(x) = 1_{[0,1]}(2x) - 1_{[0,1]}(2x - 1)$ is obtained by differencing. Whether the shape function ϕ be smooth (e.g., Gabor) or local (e.g., Haar), ψ transfers the same properties into oscillations of scaled translations of ϕ.

An essential connection to Kolmogorov structures may now be established. Recall that a Kolmogorov system is a dynamical system $(\Gamma, \mathcal{B}, \mu, S^n)$ with an order structure connecting σ-subalgebras $\mathcal{B}_n$ and a measure preserving dynamics S^n, $n \in \mathbb{Z}$ with the properties

1') $S^n \mathcal{B}_0 = \mathcal{B}_n \subseteq \mathcal{B}_m = S^m \mathcal{B}_0$, $n \leqq m$

2') $\bigcap \mathcal{B}_n = \mathcal{B}_{-\infty}$, the trivial σ-algebra of Γ

3') $\bigcup \mathcal{B}_n = \mathcal{B}$, the full σ-algebra.

The measure preserving transformations S^n induce unitary evolutions V^n on $\mathcal{L}^2(\Gamma, \mathcal{B}, \mu)$ according to Koopman's formula

$$(2.1.16) \qquad V^n f(x) = f(S^n x)$$

The Koopman construction relates trajectories in the phase space Γ to the phase functions f in the Hilbert Space $\mathcal{L}^2(\Gamma, \mathcal{B}, \mu)$. Let us denote by $\mathcal{H}_n$ and $\mathcal{H}$ the spaces $\mathcal{L}^2(\Gamma, \mathcal{B}_n, \mu)$ and $\mathcal{L}^2(\Gamma, \mathcal{B}, \mu)$ with the one-dimensional subspaces of constants subtracted off. The orthoprojections of $\mathcal{H}$ onto $\mathcal{H}_n$ are conditional expectations P_n that average the functions f in $\mathcal{H}$ with respect to the σ-algebra $\mathcal{B}_n$. The result of this nested conditional expectation ordering is the multiresolution structure

1'') $\mathcal{H}_n \subset \mathcal{H}_{n+1}$

2'') $\bigcap_n \mathcal{H}_n = \{0\}$

3'') $\bigcup_n \mathcal{H}_n$ is dense in $\mathcal{H}$.

Note that both the Kolmogorov structures and the wavelet multiresolution structures are nested chains of refinements 1), 1'), 1''). The empty intersection properties 2), 2'), 2'') imply a scale ordering for multiresolutions, a coarseness ordering for Kolmogorov systems. The completeness conditions 3), 3'), 3'') are essentially the same, the only reason the constants are subtracted off in $\mathcal{L}^2(\Gamma, \mathcal{B}, \mu)$ is because in the ergodic theory of Kolmogorov systems in statistical physics one defines the densities f as their differences from a constant equilibrium. The latter situation is in fact analogous to the regularity condition imposed on the quadrature mirror filter in the wavelet theory and also is reflected in the wavelet moment condition (2.1.15). In fact, in wavelet filter design as discussed in Daubechies [2], [3], the filter H^* which is iterated leads to all piecewise constant functions representing it converging to the shape function ϕ itself. This is a subtle but striking additional connection of the wavelet theory to ergodic theory, the theme of this section.

Property 4) of the MRA appears in many explicitly constructed Kolmogorov systems, e.g., those using Haar or Haar-like bases, for example, the iterates of the baker's transformation (1.1.13). However, Kolmogorov systems generally require the additional property (let us call it 4')) of positivity for the projections $P_n : \mathcal{H} \to \mathcal{H}_n$, as functions represent probability densities and it is desired that the K-system preserve probability densities. On the other hand positivity in K-systems implies that the projections correspond to coarse graining approximations with respect to an increasing family of σ-algebras. This property is much like the detail refinement property 4) of multiresolution analyses.

From property 4) we see that the space $\mathcal{H}_{n+1}$ is the image of the space $\mathcal{H}_n$ under the unitary transformation

$$(2.1.17) \qquad\qquad V f(x) = \sqrt{2}\ f(2x)$$

The operators V^n provide a (positivity preserving) unitary representation on the space $\mathcal{L}^2_{\mathbb{R}}$, of the discrete scaling transformations $S : x \mapsto 2x$. Formula (2.1.17) is just the (scaled) Koopman operator (2.1.16) for the scaling transformation. The properties (1–3) of the multiresolution analysis are just the conditions for the unitary transformation V to be (isomorphically) a bilateral shift in the sense of Halmos [1]. Such unitary shifts have purely absolutely continuous spectrum of uniform multiplicity, see Sz. Nagy and Foias [1]. The wavelet space

$$(2.1.18) \qquad\qquad W \equiv W_0 = \mathcal{H}_1 \ominus \mathcal{H}_0 = V\mathcal{H}_0 \ominus \mathcal{H}_0$$

is just (isomorphically) the wandering generating subspace of the shift V. The term wandering means that

$$(2.1.19) \qquad\qquad V^n W \perp V^m W, \quad n \neq m$$

and the term generating means that

$$(2.1.20) \qquad\qquad \mathcal{L}^2(\mathbb{R}) = \bigoplus_{n=-\infty}^{+\infty} V^n W$$

The wandering generating subspace $W = W_0$ is clearly infinite dimensional because the functions $\psi(x - \alpha)$, $\alpha \in \mathbb{Z}$ are an orthonormal basis for it. As the dimension of the wandering generating subspace W is the multiplicity of the shift, we conclude that the multiresolution analysis is a bilateral shift of infinite countable multiplicity.

Property 5) of the wavelet multiresolution analysis is an irreducibility requirement common to all theories related to regular representations in group theory. It is shared by both wavelet MRA and K-systems. The space H_0 of the MRA corresponds to the projection P_0 of the coarse graining induced by the generating partition ξ_0 of the K-system. Therefore we have proved

THEOREM 2.1.3.

(a) Wavelet multiresolution analyses and Kolmogorov systems possess the same basic 5 properties. The principal distinction appears in the refinement properties 4). There the wavelet MRA is generally constrained to an underlying scaling (viz., dilation, stretching) group, whereas the K-system is constrained to an underlying measure preserving group.

(b) One may define any multiresolution analysis on $\mathcal{L}_\mathbb{R}^2$ as a bilateral shift of scalings on $\mathcal{L}_\mathbb{R}^2$ with countably infinite multiplicity such that the wandering generating subspace is a cyclic representation space with respect to the group of discrete translation. The cyclic vectors are just the wavelets. This statement amounts to an operator theoretic characterization of the scaling transformations of wavelets.

REMARK 2.1.4. Theorem 2.1.3 is due to Antoniou and Gustafson [3]. Although the notion of the wavelet subspace W_0 as a generating wandering subspace is natural, we have not seen it elsewhere. In stressing the property 4) as that which may distinguish wavelet structures from Kolmogorov structures, we emphasized the underlying group properties, whereas in the discussion prior to Theorem 2.1.3, we emphasized the positivity property of K-systems. However, these are related, see our earlier discussion of the Koopman Converse Lemma 1.2.1.

REMARK 2.1.5. The continuous parameter version of Theorem 2.1.3 also holds. In fact, as will be made clear below, continuous time wavelets and K-systems are even closer to one another, through canonical commutation relations. See part (b) of Theorem 2.2.1.

2.2. Wavelets and Stochastic Processes. From Theorem 2.1.3 we see that the n-th wavelet subspace $\mathcal{W}_n$ corresponding to the detail of the multiresolution analysis at the n-th stage

$$(2.2.1) \qquad \mathcal{W}_n = \mathcal{H}_{n+1} \ominus \mathcal{H}_n = V^n W$$

is just the n-th innovation associated with the bilateral shift V. Successive innovations correspond to successive improvement of detail detection. The term innovation comes from the theory of regular stochastic processes which we will address in this section. For the convenience of the reader we repeat here the well known definition of discrete stationary regular stochastic processes. Such stationary stochastic processes in the wide sense are characterized by a discrete family of stochastic variables $\{f_n \mid n \in \mathbb{Z}\}$ over the standard Borel space X with σ-algebra $\mathfrak{S}$ and reference measure ν. The stochastic variables have finite correlation functions:

$$(2.2.2) \qquad R(f_n, f_m) = \int_x d\mu(x) f_n^*(x) f_m(x)$$

Therefore the stochastic variables belong to a Hilbert space $\mathcal{H}$ with respect to the correlation scalar product (,). Each stochastic variable f_n has therefore a finite autocorrelation:

$$(2.2.3) \qquad R(f_n) \equiv \int_x d\mu(x) |f_n(x)|^2$$

Stationarity means that the expected values $E f_n = \int_x d\mu(x) f_n$ do not depend on the variable n and that the correlation functions (,) depend only on the difference $n - m$, i.e., $R(f_n, f_m) = R(n - m)$. Regularity is expressed in terms of the family of Hilbert spaces $\mathcal{H}_{-\infty}, \mathcal{H}_n, \mathcal{H}_\infty$, $n \in \mathbb{Z}$ which are defined as follows: each $\mathcal{H}_n$, $n \in \mathbb{Z}$ is the closure of the linear span of the stochastic variables $\{f_m \mid m \leq n\}$ and represents the past and present information at the stage n. $\mathcal{H}_\infty$ is the closure of all stochastic variables $\{f_m \mid m \in \mathbb{Z}\}$ and can be identified with the Hilbert space $\mathcal{H}$ without loss of generality. The space $\mathcal{H}_\infty$ represents the remote future of the process. The space $\mathcal{H}_{-\infty}$ represents the infinite remote past and is defined by the formula $\mathcal{H}_{-\infty} = \bigcap_{n \in \mathbb{Z}} \mathcal{H}_n$.

Notice now that if we set the stationary expected values $E f_n$ equal to zero, as is commonly done in the theory of stationary stochastic processes, then the infinite remote past is empty and the stationary regular stochastic process structure is a multiresolution structure 1''), 2''), 3'').

Property 4), just as in the case of Kolmogorov systems discussed in Section 2.1, is the main distinguishing property between multiresolution structures and stationary regular stochastic process structures. A multiresolution structure in $\mathcal{L}_{\mathbb{R}}^2$ and a regular stationary stochastic process in $\mathcal{H}$ are entirely analogous in chain structure 1), 2), 3). But the former has deterministic interpretation and a precise diadic scaling stipulated while the latter has probabilistic interpretation and scaling through underlying ordered σ-subalgebras.

Property 5) is shared to a greater extent than might be at first evident. In particular, regular stationary stochastic processes when placed in a Hilbert space $\mathcal{L}^2(-\infty, \infty, \mu)$ possess the usual spectral representation and the existence of cyclic generating vectors may be established. Summarizing, we have

THEOREM 2.2.1.

(a) Stationary regular stochastic square integrable processes and wavelet multiresolution analyses possess the same basic 5 properties. The principal distinction appears in the refinement properties 4). Whereas wavelet MRA's couple a scaling group to a translation group, the stochastic processes need only underlying ordered σ-subalgebras.

(b) Any continuous time wavelet multiresolution analysis and any square integrable regular stochastic process are unitarily equivalent to a quantum mechanical momentum evolution of infinite multiplicity.

REMARK 2.2.2. Part (a) of Theorem 2.2.1 essentially says that the theory of wavelet multiresolution analyses may be placed within the theory of stationary random processes, because the latter is not encumbered with the specific group structures that make the former so interesting. But one should not regard the latter as a generalization of the former, because a wavelet MRA need not have any stochastic interpretation, and need not possess positivity preserving lattice structures inherent in stochastic process theory. Part (b) of Theorem 2.2.1 is purposely stated only for the continuous parameter case. Then the wavelet MRA properties 1), 2), 3) set up a continuous family of projections P_t which are a resolution of the identity for a selfadjoint operator. The continuous MRA property 4) becomes $f(x) \in H_t$ iff $f(\beta x) \in H_{\beta t}$, where H_s are the spectral subspaces for the operator. Property 5) is taken to mean the existence of generating cyclic vectors. Although previously we reserved the notion of MRA for the discrete case, here we brought in the continuous case so that we could directly employ the Stone–Von Neumann Lemma 1.2.7. Then part b) follows as in the proof for regular stochastic processes given in Gustafson and Misra [1]. Further discussion may be found in Antoniou and Gustafson [3].

REMARK 2.2.3. There is a large literature on regular stochastic processes, see Doob [1], Hida [1], and many later treatises. Here we are considering only square integrable processes, sometimes called those of second order. According to Walters [1], Kolmogorov originally introduced his dynamical K-systems in 1958 in analogy with regular stochastic processes. Here we are extending that analogy to wavelet multiresolution analyses.

REMARK 2.2.4. One arrives at the impression that, in spite of its roots in the underlying point or set dynamical maps, modern ergodic theory has tended to become an integration theory. See for example the book of Krengel [1], which summarizes many developments in recent ergodic theory. In particular, ergodic theory became a parallel development to the theory of martingales in stochastic process theory. Later, efforts were made to unify the (ergodic-based) integration theory and the theory of martingales. See the somewhat countervailing discussions of M. Rao [1], who highlights the differences between the two theories, and Neveu [1], who takes the view that martingale theory is contained within ergodic (integration)theory.

REMARK 2.2.5. The connections between martingales and K-systems are thus very close, although not widely stressed in the literature. Usually continuous time martingales are defined in terms of a 'filtration' $\mathcal{F}_t$ of sub-σ-fields. These are conceptually the same as our $\mathcal{B}_t$ σ-fields constructed in the proof of Theorem 1.2.6. A martingale then is defined to be a process of conditional expectations $X_t = E(X_x \mid \mathcal{F}_t)$ for all $s \geq t$. These are conceptually the same as the conditional expectations P_t constructed in Theorem 1.2.6. An $\mathcal{L}^2$ martingale which also satisfies Kolmogorov's 0–1 law is a K-system minus the underlying measure preserving requirement. A wavelet MRA is more general because its projections need not be conditional expectations.

REMARK 2.2.6. It may seem at first inexplicable that Doob [1] in his classic and pioneering development of martingale theory, chose to delve so little into the theory of the underlying trajectory maps. But the latter dynamical system theory developed later. Also, martingale theory is 'fair game' gambling theory and is usually more naturally set in $\mathcal{L}^1$ state space, with $\mathcal{L}^2$ and $\mathcal{L}^p$ versions following by modification. Even good later treatments of martingale theory such as Kopp [1] where the relationship of the martingale X_t process to the underlying filtration $\mathcal{F}_t$ (also called a stochastic basis) of increasing σ-fields is clearly exposed, do not treat the important physical role of trajectory maps S_t in the phase space. In some sense, our point of view is that the phase space mappings and the state space mappings should now be brought closer together to enrich the theory of both.

2.3. The Time Operator of Wavelets. The properties 1), 2), 3) of the multiresolution analysis imply that the approximation projections P_n satisfy the properties of a discrete resolution of the identity P_n, $n \in \mathbb{Z}$ in the Hilbert space $\mathcal{L}^2_{\mathbb{R}}$, namely

1''') $P_n < P_{n+1}$

2''') $P_{-\infty} = \lim_{n \to -\infty} P_n = 0$

3''') $P_{+\infty} = \lim_{n \to +\infty} P_n = I$

It is also straightforward to see that property 4) of the multiresolution analysis means that

(2.3.1)
$$P_{n+1} = VP_nV^{-1}$$

where V is the unitary operator (2.1.17) associated to scalings. The above properties 1$'''$), 2$'''$), 3$'''$) are the properties of a discrete system of imprimitivity, G. Mackey [1]. Therefore according to the definition of the Time Operator in statistical physics through such systems of imprimitivity, we are led to define the time operator of the multiresolution analysis by the formula

$$(2.3.2) \qquad T = \sum_{n \in \mathbb{Z}} n Q_n, \qquad Q_n = P_{n+1} - P_n$$

The imprimitivity conditions 1$'''$), 2$'''$), 3$'''$) or (2.3.1) give the change of the time operator at each stage n:

$$(2.3.3) \qquad V^{-n} T V^n = T + nI$$

The multiresolution approximation projections P_n are thus seen as the spectral projections of the time operator T. From the relation (2.1.13) of the approximation projection P_n and the wavelets $\psi_{n\alpha}$ we see that the action of the time operator in terms of wavelets is

$$(2.3.4) \qquad Tf(x) = \sum_{n \in \mathbb{Z}} \sum_{n \in \mathbb{Z}} n \langle \psi_{n\alpha}, f \rangle \psi_{n\alpha}(x)$$

The relation of the wavelets with the time operator is now clear.

THEOREM 2.3.1. Any wavelet multiresolution analysis defines a Time operator. The wavelets $\psi_{n\alpha}$ are the age eigenstates of the Time operator. Age n means detail resolution at the stage n.

Proof. Given essentially above. For more details, see Antoniou and Gustafson [2]. See Antoniou and Misra [1] for a good summary of Time operators as they are known to date in theoretical physics. See also Remarks 1.1.9 and 1.2.8. Time operators are always obtained essentially by use of the Stone–Von Neumann Lemma 1.2.7 i.e., the canonical commutation relations. Lemma 1.2.7 was stated in continuous time. The discrete parameter version is somewhat weaker but the system of imprimitivity argument used above is sufficient to obtain the Time operator for discrete wavelets. The Time operator for continuous parameter wavelets of course follows in the same way.

REMARK 2.3.2. Primas [1] attributes the notion of internal Time operator to Gustafson and Misra [1] in connection with regular stochastic processes. But Pauli [1] introduced the Time operator earlier in a quantum mechanics context. Essentially the Time operator is always a momentum evolution under unitary transformation via the canonical commutation relations.

The time operator (2.3.2) gives therefore a global description of the 1), 2), 3) properties of any wavelet multiresolution analysis. Property 4) of the multiresolution analysis is now exhibited by the time operator directly as a process of aging (scaling). Property 5) of the multiresolution analysis now is explicitly expressed spectrally as the eigenbasis of the time operator. For example, the time operator of the Haar wavelets

$$(2.3.5) \qquad \psi_{n\alpha}(x) = 2^{\frac{n}{2}}[1_{[0,1]}(2^{n+1}x - 2\alpha + 2) - 1_{[0,1]}(2^{n+1}x - 2\alpha + 1)]$$

is simply

$$(2.3.6) \qquad Tf(x) = \sum_{n\alpha} n\langle \psi_{n\alpha} \mid f \rangle \psi_{n\alpha}(x)$$

REMARK 2.3.3. A. Haar's principal tasking in his dissertation, Haar [1], was to show that there existed a complete orthonormal basis that was not derivable from any Sturm–Liouville selfadjoint differential operator. His solution was the construction of the basis (2.3.5) now named after him. We find now the natural setting of the Haar basis, namely, as the eigenbasis of the time operator for the Haar wavelet system. Theorem 2.3.1 generalizes this fact to arbitrary wavelets.

Chapter 3   Chaos in Iterative Maps

3.1. Attractors and Repellers. In one of the early computer experiments with iterated maps, Ulam and Von Neumann [1] examined the logistic map

$$(3.1.1) \qquad\qquad x_{n+1} = \beta x_n(1 - x_n), \quad 0 \leqq x_0 \leqq 1.$$

They took the parameter $\beta = 4$ so that the unit interval [0,1] was mapped onto itself at each iteration. Later May [1] emphasized its potential use for modelling population growth problems. As is well-known, recently the map (3.1.1) has received extensive attention in the literature of chaotic one-dimensional iterative mappings, see Devaney [1].

In particular, Ulam and Von Neumann [1] found that if one starts with a uniform distribution of initial conditions x_0 in [0,1], one always approaches the inverted curve

$$(3.1.2) \qquad\qquad \rho(x) = \pi^{-1} x^{-\frac{1}{2}}(1 - x)^{-\frac{1}{2}}$$

in the limit as $n \to \infty$, in the following sense: $\rho(x)$ is the fraction of the time the iterations spend at the value x. Generally for iterated maps such $\rho(x)$ are now called the *natural invariant density* on the attractor [0,1]. Periodic orbits are dense in [0,1], but they are countable so their probability of occurrence is zero. Nonperiodic orbits are therefore the typical behavior which generates the natural invariant density (3.1.2).

Chaos has many definitions but here it may be taken as: exponential divergence of orbits from close initial conditions. This is equivalent to a positive Lyapunov exponent which represents the rate of divergence of orbits. Thus chaotic deterministic dynamical systems are those which violate the third tenet of well-posed problems, that of stability with respect to initial conditions.

The invariant density (3.1.2) is singular at $x = 0, 1$. Generally for iterated maps one can expect singular behavior to appear at unknown locations. To cope with this, one generalizes the notion of natural invariant density to the notion of *natural invariant measure* as follows. Consider an iterated map (let us restrict it to the interval [0,1] for simplicity here)

$$(3.1.3) \qquad\qquad x_{n+1} = Sx_n$$

and consider any probability measure μ (nonnegative countably additive set function on [0,1], $\mu[0, 1] = 1$). Then μ is called invariant if $\mu(S^{-1}\omega) = \mu(\omega)$ for any measureable subset ω. Here

$S^{-1}\omega$ means all points which mapped to ω during an iteration. Given any interval $[a, b]$ in $[0,1]$, let $\mu([a, b], x_0)$ be the fraction of time spent in $[a, b]$ by an orbit emanating from any initial point x_0 in $[0,1]$, in the limit as $n \to \infty$. Assume that these limiting values $\mu([a, b], x_0)$ exist for all x_0 in $[0,1]$ except perhaps for a set $\sum$ of Lebesgue measure zero, and that they are the same. Then μ is called the *natural invariant measure* on $[0,1]$. The same definition applies if $[0,1]$ is replaced by x_0 in the basin of attraction of any attractor set. For the logistic map (3.1.1) with $\beta = 4$, $d\mu(x) = \rho(x)dx$ with ρ from (3.1.2), so the natural invariant measure is just $\mu[a, b] = \int_a^b \rho(x)dx$.

This construction of natural invariant measure can be taken on any specified subset Ω of $[0,1]$ and more generally for any subset of interest (e.g., a fractal attractor) within the domain of any iterated map $x_{x+1} = Sx_n$. It has been shown, Sinai [1], Ruelle [1] and Bowen and Ruelle [1] that a natural invariant measure exists for all Axiom A attractors: those of hyperbolic dissipative dynamical systems for which the attractor possesses a dense set of periodic orbits.

REMARK 3.1.1. The baker's transformation (1.1.13) is an example of an Axiom A attractor. Hyperbolic dynamical systems are those, roughly, whose linearizations at each point on the attractor possess no purely imaginary eigenvalues. For example, their fixed points are structurally stable, in the sense that their qualitative features remain the same under small perturbations. Another way to view hyperbolicity is that at all points on the attractor, one may decompose the flow along stable manifolds and unstable manifolds with the absence of any center manifold. For the baker's map, the unstable manifolds are horizontal lines (stretching directions) and the stable manifolds are vertical lines (flattening directions). The attractor is the whole unit square. Area is preserved and the baker's transformation is a deterministic Kolmogorov-system in the terms of Chapters 1 and 2. It possesses a countable number of periodic orbits, and an uncountable number of aperiodic orbits.

If the baker's transformation is varied to

$$(3.1.4) \qquad (x_{n+1}, y_{n+1}) = S(x_n, y_n) = \begin{cases} (2x_n, ay_n) & 0 \leq x < 1/2 \\ (2x - 1, ay_n + 1/2) & 1/2 \leq x \leq 1 \end{cases}$$

the area preserving property is lost. If $a < 1/2$, the attractor changes from the whole square to a fractal Cantor set of horizontal lines. This area shrinking in phase space is called *dissipation*. On the other hand, if $a > 1/2$, some points are expelled out of the top of the square. These may just be cut off, or may be regarded as going to the attractor point $y = \infty$. Such maps often possess fractal repellers which may be analyzed similarly to attractors. Often the points which remain nonrepelled exhibit very interesting chaotic trajectories before they finally fall onto an attractor, or exhibit some other kind of long time behavior influenced by both attractors and repellers within the system. Examples of repellers are unstable attractors, e.g., unstable fixed points or unstable periodic orbits.

The latter dynamical systems, in which some points escape the basic phase space, and in which one finds fractal repellers or almost attractors, have been less studied, although in the last fifteen years a literature has accumulated. Often they are considered within the frame of dissipative dynamical systems, but it would seem that they may be regarded as a class of dynamical systems distinct from dissipative or conservative systems. One might suggest the term *accretive* systems for them, in analogy with that term as used in the operator semigroup theory. Whereas dissipative systems lose volume, accretive systems gain volume, at least near their repeller sets. Of course when these mapped volumes meet the boundary of the phase space, a portion is ejected. We will study some one-dimensional accretive systems in the next section.

REMARK 3.1.2. The principle of probability preference of Section 1.3, in particular its violation by loss of probability density function, is illustrated by these simple one-dimensional accretive systems to be studied in the next section. In the quantum microscopic irreversibility model discussed in Section 1.3, the logistic map (3.1.1) and its continuous counterpart the sigmoid map were employed. The logistic map (3.1.1) is neither dissipative nor accretive: it expands or contracts intervals in the phase space depending on whether you are near the ends or center of the unit interval, respectively. This property is perhaps why it exhibits such an interesting behavior, some aspects of which are still not understood. For $\beta > 4$ a phase space subinterval symmetric about $x = 1/2$ is expelled out of phase space, even though you are in a dissipative subdomain of the map. This is an example of probability violation as discussed in Section 1.3.

REMARK 3.1.3. At risk of saying the obvious, we may editorially point out that in probability theory, one becomes inured to the idea that probability density functions and their cumulative probability distribution functions are one to one. But this is not the case, the derivatives of the latter no longer give the former if a slope exceeds one.

3.2. A Gap Theory for Information Dimension. In this section we turn our attention to

some simple one-dimensional accretive maps, and calculate their information dimension D_1. These maps are generalizations of the simple Renyi map

$$(3.2.1) \qquad x_{n+1} = Sx_n = 2x_n \ (\text{mod}\,1)$$

on the unit interval. This is a very chaotic system. Its action may be viewed as one of stretch, twist, and fold on a torus. Each iteration chaotically separates nearby points by a factor of two. Points in the interval on periodic orbits are dense but countable and thus nontypical. The natural invariant density is $\rho(x) = 1$ on $[0,1]$, the maximal invariant set is the full interval $[0,1]$, the natural invariant measure is Lebesgue measure.

REMARK 3.2.1. The Renyi map S is an example of an exact endomorphism. Given a measure space $(\Omega, \mathcal{B}_0, \mu)$ and a map S, σ-subalgebras are defined by

$$(3.2.2) \qquad \mathcal{B}_{-n} = S^{-n}\mathcal{B}_0$$

i.e., they become more and more coarse. If $\bigcap_n \mathcal{B}_{-n}$ is the trivial σ-algebra, i.e., just the subsets of $\mathcal{B}_0$ which have measure 0 or 1, the map S is called an exact endomorphism. Note that the latter condition on $\bigcap_{-n} \mathcal{B}_{-n}$ is the condition (3) of (1.1.14) of Chapter 1, the "emptyness of the infinite remote past" of Kolmogorov Systems. From this one induces the following.

THEOREM 3.2.2. Exact systems have natural extensions to Kolmogorov Systems.

Proof. See Cornfeld, Fomin and Sinai [1]. The natural extension of the Renyi map (3.2.1) is the baker's map (1.1.13). Thus exact chaotic systems can be embedded within conservative chaotic systems.

Next let us recall the important notion of information dimension D_1. This dimension is one of the most interesting fractal dimensions because it accounts for the relative frequencies with which orbits visit points or infinitesimal regions in the phase space. The corresponding natural invariant measures for such systems will in general be singular. The nature of those singularities, e.g., spike-like or delta-function-like, contains much more information about the dynamical system than does just the attractor, repeller, or other fractal set upon which the measure has support.

REMARK 3.2.3. There is too much to say about fractal dimension theory to summarize here. We refer the reader to Mandelbrot [1], Falconer [1] for excellent accounts. The usual fractal dimension is that of box-counting, or what is often the same, Hausdorff, dimension. The fractal set of interest is covered with small boxes of side-length ϵ. Let $N(\epsilon)$ be the minimum number of N-cubes of side-length ϵ needed to cover the set. Then

$$(3.2.3) \qquad d = \lim_{\epsilon \to 0} \frac{\ln N(\epsilon)}{-\ln \epsilon}$$

is the fractal dimension. But note that this in no way measures the dynamics on the fractal set, i.e., the frequency of visits of the dynamics to particular portions of the set. Therefore consider an attractor, or other invariant set of interest. Let

$$(3.2.4) \qquad D_q = \frac{1}{1-q} \lim_{\epsilon \to 0} \frac{\ln \sum_{i=1}^{N(\epsilon)} (\mu_i)^q}{-\ln \epsilon}$$

where $N(\epsilon)$ is the number of ϵ-cubes in the cover, q is a real number, and μ_i is the natural measure of each cube in the cover, namely

$$(3.2.5) \qquad \mu_i = \lim_{T \to \infty} \frac{\text{time in } i\text{th cube}}{T}$$

for any typical orbit starting from a point x_0 which is attracted to the box. The information dimension is defined as

$$(3.2.6) \qquad D_1 = \lim_{q \to 1} D_q$$

THEOREM 3.2.4. Generally the information dimension is given by

$$(3.2.7) \qquad D_1 = \lim_{\epsilon \to 0} \frac{\sum_{i=1}^{N(\epsilon)} \mu_i \ln \mu_i}{\ln \epsilon}$$

The dimensions D_q decrease with increasing q, for example,

$$(3.2.8) \qquad D_2 \leqq D_1 \leqq D_0$$

D_0 is the usual box-counting dimension.

Proof. See Falconer [1] and the references therein.

Let us now consider a particular set of maps of the unit interval

$$(3.2.9) \qquad x_{n+1} = S x_n = \begin{cases} \beta x_n & 0 \leqq x_n \leqq \beta^{-1} \\ \beta' x_n + (1 - \beta') & 1 - (\beta')^{-1} \leqq x_n \leqq 1 \end{cases}$$

These are variations of the Renyi map (3.2.1) where now one has different slopes on the two pieces of the map. If the map exceeds the value one, we regard the orbit as having escaped, as mentioned above when we introduced the notion of accretive dynamical systems. We will restrict attention in the rest of this section to maps (3.2.9). A generic example, which we will use to illustrate the results, is shown in Fig. II-2. There we took $\beta = 2$ and $\beta' = 5$. A well known example is the singular Cantor map $\beta = \beta' = 3$. We shall always assume

$$(3.2.10) \qquad\qquad \beta > 1, \quad \beta' > 1, \quad \beta^{-1} + (\beta')^{-1} \leqq 1$$

The maximal invariant set is [0,1] when $\beta^{-1} + (\beta')^{-1} = 1$, and otherwise a Cantor-like fractal set with fractal dimension D_0 satisfying $\beta^{-1} + ((\beta')^{D_0})^{-1} = 1$. When $\beta \neq \beta'$, the latter invariant set is not uniformly scaled, and is called a multifractal.

THEOREM 3.2.5. Any map (3.2.8) possesses an uncountable number of ergodic invariant measures on its maximal invariant set. These are in general singular, and hence not SRB (Sinai–Ruelle–Bowen) measures.

Proof. An invariant measure μ will be obtained via its distribution function $F(x) \equiv \mu(0, x]$ if F satisfies

$$(3.2.11) \qquad F(x) = F\left(\frac{x}{\beta}\right) + F\left(\frac{x + \beta' - 1}{\beta'}\right) - F\left(1 - \frac{1}{\beta'}\right)$$

because in terms of the measure μ, this is the same as

$$(3.2.12) \qquad \mu(S^{-1}(0, x]) = \mu(0, x/\beta] + \mu(1 - 1/\beta', x/\beta' + 1 - 1/\beta'].$$

To that end, introduce the following operator T_α and free parameter $0 < \alpha < 1$

$$(3.2.13) \qquad T_\alpha F(x) = \begin{cases} \alpha F(\beta x) & 0 \leq x \leqq 1/\beta \\ \alpha & 1/\beta \leqq x < 1 - 1/\beta' \\ (1 - \alpha)F(\beta' x + 1 - \beta') + \alpha & 1 - 1/\beta' \leqq x \leqq 1 \end{cases}$$

Following De Rham [2], for each $0 < \alpha < 1$ there is a unique fixed point $F_\alpha(x)$ of the equation $T_\alpha F(x) = F(x)$. This follows by verifying that T_α is a strictly contractive map on $\mathcal{L}^\infty[0, 1]$. Putting $F_\alpha(x)$ into the three portions of (3.2.12) gives

$$\begin{aligned} & F_\alpha\left(\frac{x}{\beta}\right) + F_\alpha\left(\frac{x + \beta' - 1}{\beta'}\right) - F_\alpha\left(1 - \frac{1}{\beta'}\right) \\ (3.2.14) \qquad & = \alpha F_\alpha(x) + (1 - \alpha)F_\alpha(x) + \alpha - \alpha \\ & = F_\alpha(x) \end{aligned}$$

and hence the condition (3.2.10) for the invariance of the measure μ is satisfied. By the De Rham construction, generally $F_\alpha(x)$ is a nonnegative nondecreasing continuous function with zero derivatives a.e. Since SRB measures are absolutely continuous with respect to Lebesgue measures, the induced measures here are not SRB measures.

REMARK 3.2.6. Theorem 3.2.5 was shown in Tasaki, Suchanecki and Antoniou [1]. For a good discussion of SRB measures, see Eckmann and Ruelle [1].

It turns out that for the maps (3.2.9), the information dimension D_1 of the induced singular invariant measures $\mu_{\alpha,\beta,\beta'}$ is given by the expression.

$$
D_{\alpha,\beta,\beta'} = \frac{h}{\lambda} = \frac{\text{Kolmogorov–Sinai Entropy}}{\text{Lyapunov Exponent}}
$$

(3.2.15)

$$
= -\frac{\alpha \ln \alpha + (1-\alpha)\ln(1-\alpha)}{\alpha \ln \beta + (1-\alpha)\ln \beta'}
$$

The following results for the information dimension for these maps, and n-piece generalizations of them, were obtained in Gustafson [6].

LEMMA 3.2.7. (upper bounds). For arbitrary $\beta > 1$, $\beta' > 1$, $1/\beta + 1/\beta' \leqq 1$, $0 < \alpha < 1$, the information dimension $D_{\alpha,\beta,\beta'}$ has upper bounds

(a)
$$
D_{\alpha,\beta,\beta'} \leqq 1 - \left[\frac{1 - (\beta^{-1} + (\beta')^{-1})}{\alpha \ln \beta + (1-\alpha)\ln \beta'}\right]
$$

(b)
$$
D_{\alpha,\beta,\beta'} \leqq \frac{1}{1 + \left[\frac{1-(\beta^{-1}+(\beta')^{-1})}{-\alpha \ln \alpha - (1-\alpha)\ln(1-\alpha)}\right]}
$$

The second upper bound is sharper than the first.

Proof. (a) We may write

$$
D_{\alpha,\beta,\beta'} \equiv \frac{-\alpha \ln \alpha - (1-\alpha)\ln(1-\alpha)}{\alpha \ln \beta + (1-\alpha)\ln \beta'}
$$

(3.2.16)
$$
= \frac{(\alpha - \alpha \ln \alpha) + ((1-\alpha) - (1-\alpha)\ln(1-\alpha)) - 1}{\alpha \ln \beta + (1-\alpha)\ln \beta'}
$$

$$
\leqq \frac{\beta^{-1} - \alpha \ln \beta^{-1} + (\beta')^{-1} - (1-\alpha)\ln(\beta')^{-1} - 1}{\alpha \ln \beta + (1-\alpha)\ln \beta'}
$$

where we have twice used the Gibb's inequality, $a - a\ln a \leqq b - a\ln b$, a and b positive entities.

(b) In like manner, but now focusing attention on the denominator rather than the numerator, we have

$$
\alpha \ln \beta + (1-\alpha)\ln \beta' = \beta^{-1} - \alpha \ln(\beta^{-1}) + (\beta')^{-1} - (1-\alpha)\ln(\beta')^{-1} - (\beta^{-1} + (\beta')^{-1})
$$

$$
\geqq \alpha - \alpha \ln \alpha + (1-\alpha) - (1-\alpha)\ln(1-\alpha) - (\beta^{-1} + (\beta')^{-1})
$$

$$
= -\alpha \ln \alpha - (1-\alpha) - (1-\alpha)\ln(1-\alpha) + [1 - (\beta^{-1} + (\beta')^{-1})]
$$

from which the second upper bound follows.

That the second upper bound in Lemma 3.2.7 is sharper than the first upper bound will follow trivially from the Lemma below, but let us directly prove the stated relationship here. For that and later purposes, it is convenient to call the quantity $[1 - (\beta^{-1} + (\beta')^{-1})]$ the "Gap." Considering the first upper bound minus the second upper bound, we have (in obvious shorthand)

$$(3.2.17) \qquad \begin{aligned} \left(1 - \frac{\text{gap}}{\text{denomin}}\right) - \left(\frac{\text{numerator}}{\text{numer} + \text{gap}}\right) &= \frac{d - g}{d} - \frac{n}{n + g} \\ &= \frac{(d - g)(n + g) - nd}{d(n + g)} = \frac{dg - g^2 - gn}{d(n + g)} \\ &= \frac{g}{d(n + g)} \left[d - (n + g)\right] \end{aligned}$$

Thus the second upper bound $\leq$ the first upper bound if and only if $d \geq n + g$. This may be verified directly, e.g., as in the proofs of (a) and (b) above.

More to the point is the following interesting lemma, which is sharp. For the lemma and later purposes, it is now convenient to introduce and discuss the notations:

g, Gap, information gap, Cantor Gap, $[1 - (\beta^{-1} + (\beta')^{-1})]$, more generally, $1 - \sum_{i=1}^{n} q_i$, where each $a_i > 0$ and $\sum_{i=1}^{n} q_i = q \leqq 1$. Gap represents "lost information" in the sense of failure of base interval matchup for the original map S. Although we won't detail the more general (n pieces) maps S which could be associated with the more general gap formulation just given, the gap $1 - q$ measures the failure of the slopes of the inverse map S^{-1}, as interpreted as probabilities, to be optimal. It also measures the total "Cantor effect" of interval removal caused by the repellers of the invariant set of the map S. It also corresponds to the amount of probability violation in the sense of Section 1.3.

h, Entropy, H_{metric}, Kolmogorov–Sinai Entropy, $-\alpha \ln \alpha - (1 - \alpha) \ln(1 - alpha)$, more generally, $-\sum_{i=1}^{n} p_i \ln p_i$, where each $p_i > 0$ and $\sum_{i=1}^{n} p_i = 1$. We could also introduce an incomplete entropy here, where the p_i do not sum to one, which would introduce an "entropy gap" similar to the "information gap" just discussed, however, for simplicity we will not do so. The roles of the p_i will be similar to that of α and $1 - \alpha$, notably, α may be thought of as the "slope" of S^{-1} with respect to the measure μ_α, $dS/d\mu_\alpha = \alpha^{-1}$. The most important operational role of α in these theories is to preserve measure additivity as expressed in the distribution function $F_\alpha(x)$

$$(3.2.18) \qquad F_\alpha(x) = \begin{cases} \alpha F_\alpha(\beta x) & 0 \leq x < \beta^{-1} \\ \alpha & \beta^{-1} \leq x < 1 - (\beta')^{-1} \\ (1-\alpha)F_\alpha(\beta' x + 1 - \beta') + \alpha & 1 - (\beta')^{-1} \leq x \leq 1 \end{cases}$$

which solves the De Rham equation as shown above. It is by this decomposition that the distribution function $F_\alpha(x)$ satisfies the De Rham equation. The placement of the value α in the Gap g preserves the additivity of the measure μ_α.

λ Lyapunov exponent, $\alpha \ln \beta + (1-\alpha) \ln \beta'$, more generally, $\sum_{i=1}^{n} p_i \ln q_i$, where the p_i and q_i were defined above. Lyapunov exponents are of course related to decay rates and trajectory divergences and I will return to this point below.

G, Gibbs discrepancy, $(\beta^{-1} - \alpha \ln(\beta^{-1}) + (\beta')^{-1} - (1-\alpha)\ln(\beta')^{-1}) - (\alpha - \alpha \ln \alpha + (1-\alpha) - (1-\alpha)\ln(1-\alpha))$, more generally, $\sum_{i=1}^{n}[(q_i - p_i \ln q_i) - (p_i - p_i \ln q_i)]$, where the p_i and q_i were defined above. This is a new notion growing out of the proof of the upper bounds of Lemma 3.2.7 above. Its key role is apparent in the following lemma, which sharpens those bounds.

LEMMA 3.2.8. (Lyapunov decomposition).

$$\lambda = g + G + h.$$

Proof. For the special case, we may write

(3.2.19)
$$\alpha \ln \beta + (1-\alpha)\ln \beta' = -[\beta^{-1} + (\beta')^{-1}] + \{[\beta^{-1} - \alpha \ln(\beta^{-1}) + (\beta')^{-1} - (1-\alpha)\ln(\beta')^{-1}]$$
$$- [\alpha - \alpha \ln \alpha + (1-\alpha) - (1-\alpha)\ln(1-\alpha)]\} + \{numer + 1\}.$$

For the general case, the argument is the same: in shorthand, in terms of the numerator and denominator of the information dimension $D = numer/denom$,

$$\text{denom} = -\sum_{i=1}^{n} q_i + G + (numer + 1)$$
$$= g + G + h$$

THEOREM 3.2.9. (Information optimality). $D_{\alpha,\beta,\beta'}$, or more generally, $D_{p_1,\ldots,p_n,q_1,\ldots,q_n}$, may be expressed as

$$D = \frac{h}{g + G + h}.$$

Proof. Lemma 3.2.7.

Note that the requirements for D to be optimal ($D = 1$), even in the many parameter case, are quite evident from the theorem: if and only if $G = g = 0$. Moreover, the upper bounds of Lemma 3.2.7 are now immediate from the theorem. The first upper bound means

$$D = \frac{h}{g + G + h} \overset{\leq}{=} \frac{h + G}{g + G + h}$$

and the second upper bound means

$$D = \frac{h}{g + G + h} \overset{\leq}{=} \frac{h}{g + h}.$$

The error of the first upper bound is caused by the presence of the Gibbs discrepancy in the numerator. The error of the second upper bound is caused by the absence of the Gibbs discrepancy in the denominator. The right way to think of the departure from optimality, from this information dimension view, is

(3.2.20)
$$D = \frac{1}{1 + \frac{G}{h} + \frac{g}{h}}$$

REMARK 3.2.10. Clearly versions of Lemmas 3.2.7, 3.2.8, and Theorem 3.2.9 also hold with the finite sums replaced by convergent infinite sums or integrals, so that a countable or uncountable infinity of parameters can be accommodated. Other upper bounds are obtainable from D as expressed in the theorem. Lower bounds (poor) may be obtained for example by deleting specific numerator terms.

Next let us turn to the suboptimal case: $\beta^{-1} + (\beta')^{-1} = q < 1$. For fixed β, β', what should α be, to maximize $D_{\alpha,\beta,\beta'}$? With an eye toward further generalization, let us first derive here the exact α value at which D_α is maximized. For that purpose, let $x = \alpha$, $y = \beta^{-1}$, $z = (\beta')^{-1}$, $a = \ln y$, $b = \ln z$, $c = \ln(y/z)$, and let us also use, as above, the shorthand n and d for numerator and denominator of D. Then, writing out all terms,

$$\frac{\partial D}{bx} = d \ln \left(\frac{c}{1-x} \right) - n \ln \left(\frac{y}{z} \right)$$

$$= ax \ln x - ax \ln(1-x) + b \ln x - bx \ln x - b \ln(1-x) + bx \ln(1-x)$$

(3.2.21)
$$- cx \ln x - c \ln(1-x) + cx \ln(1-x)$$

$$= (a - b - c)x \ln x + (-a + b + c)x \ln(1-x) + (b) \ln x + (-b - c) \ln(1-x)$$

$$= b \ln x - a \ln(1-x)$$

where we have made use of the fortuitous cancellation $c = a - b$. Thus $\partial D/\partial x = 0$ if and only if x satisfies the (generally transcendental) equation

$$x^{\ln \beta'} = (1 - x)^{\ln \beta}$$

As the exponents are both positive, the one parameter optimal solution $x = \alpha$ is exactly the intersection of two power curves. The maximized value of D there has the interesting general expression

(3.2.22)
$$D = \frac{\ln[\alpha^\alpha (1 - \alpha)^{(1-\alpha)}]}{(a - b)\alpha + 1}$$

$$= \frac{\left[1 + \left(\frac{\ln \beta'}{\ln \beta}\right)\alpha^{\left(\frac{\ln \beta'}{\ln \beta}\right)}\right]}{\left[1 + \left(\ln(\frac{\beta'}{\beta})\right)\alpha\right]} \cdot \ln \alpha$$

For the example shown in Fig. II-2, where $\beta = 2$, $\beta' = 5$, the equation for α becomes (approximately)

$$x^{2.3219287} + x - 1 = 0$$

from which $x \approx 0.6425$ and $D \approx 0.63873$.

The second derivative is also easily found to be

$$\frac{\partial^2 D}{\partial x^2} = \ln[z^{1/x} y^{1/1-x}] = \ln\left[\frac{1}{(\beta')^{1/x}(\beta)^{1/1-x}}\right]$$

which proves the strict concavity downward for all the $D(\alpha)$ curves.

Now, to generalize the above considerations to the three parameter view, we may employ Lagrange multipliers. We wish to find the extrema of

$$w = D + \lambda(y + z - q).$$

We form the system

(3.2.23)
$$w_x = b \ln x - a \ln(1 - x) = 0$$
$$w_y = \frac{n}{d^2}\left(\frac{x}{y}\right) + \lambda = 0$$
$$w_z = \frac{n}{d^2}\left(\frac{1 - x}{z}\right) + \lambda = 0$$
$$w_\lambda = y + z - q = 0.$$

From the first equation we have $x^{\ln \beta'}$ again as above, from the second and third equations we have $z/y = (1-x)/x$, the fourth equation just being the gap constraint equation. The first equation is actually oversimplified, before cancellations it may be written

$$d \ln \left(\frac{1-x}{x} \right) - n \ln \left(\frac{z}{y} \right) = 0$$

which now tells us that for an extremum we must necessarily have

$$\left(\frac{1-x}{x} \right)^{d-n} = 1$$

There are now two cases. First, when $d = n$, any x will suffice. But that is the optimal case $D = 1$, and we know from Theorem 3.2.9 that then the gap $g = 0$ (so $\beta^{-1} + (\beta')^{-1}$ necessarily). Second, when $d \neq n$, necessarily $1 - x = x$ so $\alpha = 1/2$, and $z = y$, so $\beta = \beta' = 2/q$. The three parameter relative maximum is thus

$$D = \frac{\ln 2}{\ln 2 - \ln q}$$

The $n + 1$ parameter case can be similarly treated, in which the $\ln 2$ will change to $\ln n$.

For the example of Fig. II-2, we found above that $\alpha = 0.6425$ was maximizing for $\beta = 2$ and $\beta' = 5$, and then $D = 0.63873$. The above considerations show that for $\beta^{-1} + (\beta')^{-1} = q = 0.7$ fixed, maximal D is 0.66025 is achieved at $\alpha = 0.5$, $\beta = \beta' = 2.857143$. In principle then, Lagrange multipliers will enable exact maximization of information dimension in suboptimal cases for n parameters. Second derivative tests should also be employed to guarantee downward concavity at the local extrema, and of course behavior at the boundaries of the constraint regions should also be checked, as insurance.

We close this section with a discussion of chaotic transients, following Kantz and Grassberger [1] and Gustafson [6]. When fractal repellers are present, there occur longlived chaotic transients until the orbit is finally either expelled from the phase space or drifts eventually toward an attractor. From this one finds a recently developing chaotic scattering theory.

Let α (no relationship to the parameter α above, but we follow convention) denote the decay rate of a longlived chaotic transient. That is, $e^{-\alpha}$ is the probability that a uniformly distributed x_0 remains in the $[0,1]$ interval. This relaxation constant α is generally independent of initial distribution, provided that it was smooth. In terms of information dimension D and Lyapunov exponent λ, one has for a large class of maps, including those considered above, the following important relations to decay rate α and entropy h.

THEOREM 3.2.11.

(3.2.24)
$$\alpha = (1 - D)\lambda$$
$$\lambda = h + \alpha$$

Proof. See Kantz and Grassberger [1], Bohr and Rand [1].

The heuristic interpretation of the second relation is interesting: from a flux λ of incoming digits (some significant, some not), the fraction h/λ leads to unpredictable motion on the repeller, the fraction α/λ pushes the trajectories away from the repeller. The following extends that interpretation.

LEMMA 3.2.12. (Decay components).

(3.2.25)
$$\alpha = g + G$$

Proof. Lemma 3.2.8 and Theorem 3.2.11. The fraction α/λ in fact decomposes to

(3.2.26)
$$1 = \frac{h}{\lambda} + \frac{g}{\lambda} + \frac{a}{\lambda},$$

so that the pushing away of trajectories from the repeller is caused by two agents: the information loss, i.e., Cantor extraction loss, due to the failure of the β, β', more generally the q_i, to map the whole interval; and the Gibbs' loss, due to the relative incompatibility of the p_i to the q_i.

REMARK 3.2.13. See Gustafson [6] for further implications of these results toward conjectures of Frederickson, Kaplan, Yorke and Yorke [1] and Kadanoff and Tang [1]. An advantage of the gap theory for information dimension presented here is that one obtains values or estimates for the decay rate α in terms of the information gap g, the Gibbs gap G, and an entropy gap γ should one be present, without calculating or estimating Jacobian derivatives and without necessarily assuming total hyperbolicity of the dynamical system.

REMARK 3.2.14. An interpretation of the maximization of information dimension as an instance of the principle of regularization of Section 1.3 may be found in Gustafson [6], [8].

REMARK 3.2.15. The original motivation of De Rham [2] is enlightening. De Rham was watching M. Andre Ammann, a student in l'Ecole des Arts et Metiers in Geneva, Switzerland, make a round broom handle by starting with a beam of square cross-section, the then beveling the corners successively. The student asked De Rham what the limit of these successive bevelings would be. De Rham's answer as we know is: a continuous but generally nondifferentiable curve. This story may be found in De Rham [1].

The final 'broom handle' of the De Rham scenario epitomizes The Principle of Refinement of Detail: a final equilibrium solution, macroscopically very regularized (seemingly a circle) according to the principle of regularization, and corresponding to a physical grinding down process, yet at the microscopic scales the curve exhibits great detail, corresponding to the work of beveling that went into its formation.

3.3. Onset of Chaos in Neural Learning. The Backpropagation algorithm is perhaps the most widely employed of all neural net learning algorithms. Backpropagation owes its success to its robustness: it finds solutions in a wide variety of situations and with considerable tolerance for the setting of architecture, parameters, and training data. In this section a fundamental explanation for that robustness is proposed. A related question, why it is preferable to start with small weights, is addressed. Intermediate and final dynamics are explained in the same way. These results are taken from Gustafson [9] and originated from an earlier consideration of a sigmoid calculus, Gustafson [5]. Only a sketch of this new theory will be given here. The basic idea is that the weight updates in the Backpropagation algorithm will exhibit logistic map chaotic behavior when what we call effective dynamic gain becomes relatively large. Most neural learning algorithms have employed sigmoidal thresholding to provide nonlinear decision transitions, but to date a fundamental explanation of the dynamics and robustness of these algorithms has been lacking.

Let us consider the standard backpropagation weight increment rule (see, e.g., Hertz, Krogh, Palmer [1]) at an output node

$$(3.3.1) \qquad \Delta w_{k_j} = \eta f'(\text{net})(t_k - o_k)o_j$$

where η is the assigned learning parameter, t_k is the learning target, o_k is the net output at the current iteration, and o_j is the transmitting node value. For simplicity let us take binary input and target values, $t_k = 1$ to signify a yes output to be learned, $t_k = 0$ to signify a no output to be learned. We shall take f to be the standard sigmoid

$$(3.3.2) \qquad f(x) = \frac{1}{1 + e^{-\beta x}} = \frac{1}{2} + \frac{\beta}{4} x - \frac{\beta^3}{8} x^3 + O((\beta x)^5).$$

Here β is the assigned gain parameter, $\beta > 0$. Note that we have expanded $f(x)$ about $f(0) = \frac{1}{2}$, which will be seen to be advantageous. We are thinking about an output layer node k and we want the net response o_k to build upward or downward toward the value t_k but the same considerations will apply with modification to any hidden layer node as well. For simplicity we have taken the sigmoid bias θ to be zero, even though as was pointed out in Gustafson [5] it is important generally to allow the squashing function bias to vary, because then its weight is seen to be the most sensitive, most important, of the weights; that being however an issue separate from the thrust of this paper. In the above, net $= \sum w_j o_j$. If for simplicity we assume a single hidden layer, then if x_i is the ith component of a binary input vector, the output o_j of the jth hidden layer node is $o_j = f_j(\sum w_{ji} x_i)$, where f_j is another sigmoid.

For simplification of notation, let us drop the output node subscript k. Let us focus attention on the thresholding dynamics at this output node. For simplicity, let us consider only the case $t = 1$; as is well known, the same considerations can be modified to treat $t = 0$. The weight changes to this node are

$$(3.3.3) \qquad \Delta w_j = \eta f' \left[\sum_j w_j f_j \left(\sum_i w_{ji} x_i \right) \right] f_j \left(\sum_i w_{ji} x_i \right) (t - o).$$

To fix ideas, we now want to formally decouple the sigmoid actions so that we gain insight into just the final thresholding dynamics. A convenient way to do this is to just regard o_j as an accumulation, let us call it x_j, of the f_j thresholded weighted inputs $w_{ji} x_i$ coming into hidden layer node j. Moreover, we will then just denote the sum $\sum_j w_j x_j$ by wx and regard it as an aggregate variable for the ensuing analysis.

Of course the successive Backpropagation iterations will continually modify these internal aggregate variables but as we stated at the beginning, we want to concentrate on establishing a fundamental connection between individual sigmoid thresholding dynamics in Backpropagation and the quadratic map orbit behaviors from dynamical systems theory. The more complex, coupled, sigmoid dynamics of the full Backpropagation dynamical system may be studied later as a more general problem.

Let us look now at the initial dynamics, from a small starting weight $w = w_0$ in our initial aggregate weight wx, x representing the aggregate weight of input. We then update the weight to

$$w_1 = w_0 + \Delta w = w + \eta f'(wx)(1 - o)$$

(3.3.4)
$$\cong w + \eta(1 - o)\frac{\beta}{4}\left[1 - \frac{\beta^2 x^2}{4} \, w^2\right]$$

$$= \frac{\beta\eta(1 - o)}{4} + w\left[1 - \frac{\beta^3 x^2 \eta(1 - o)}{16} \, w\right].$$

All terms $O(x^4)$ have been and are to be dropped, although retaining them would not change the conclusions of this paper. The first term above is a residual error, to be driven toward zero. Note that, given small initial weights w, right away this first term dominates; the w and w^2 in the second term being small means that the first term sets the sign of the weight. Because o is initially zero, this starts the iteration in the (correct) direction toward $t = 1$.

The second term above may be written

$$w - \frac{\beta\eta(1 - o)}{4} \cdot \frac{\beta^2 x^2}{4} \cdot w^2$$

and we see the appearance of the residual error first term now entering as a factor in the second term. Thinking first of input x as small (think of it as $x = 1$), when initial w is small we see that the chances of this second term overcoming the weight sign set by the residual error first term are small. Thus the weight grows positively and o can be expected to grow toward the desired target value of 1.

In the other instance of $t = 0$, with o initially zero, from the above expressions we see that the first weight change is just $w_1 = w_0$. With w_0 small, the node remains "undecided". However, in our decoupled analysis above, with all initial weights assumed small and aggregate input x also small, we may expect the output $o \approx 1/2$ and the aggregate thresholded $x_j \approx o_j$ to be near $1/2$, from which the second weight change $\Delta w_j \approx \eta f'(wx)(-o)(1/2) \approx O(-1/2)$. This negative weight increment will dominate any assigned small initial weight w_0 whatever its sign. Thus the weight values start moving in the negative direction and hence the iterated values of o move thereafter in the (correct) direction toward $t = 0$.

In terms of our truncated $O((\beta x)^5)$ power series above, the above paragraph's discussion becomes

(3.3.5)
$$w_2 \approx -\frac{\beta\eta}{8} + w_0\left[1 + \frac{\beta^3 x^2 \eta w_0}{32}\right]$$

For example with $\beta = \eta = x = 1$ this becomes $w_2 = -1/8 + w_0(1 + w_0/32)$ which for small w_0 goes in the (correct) negative weight direction.

Had we used a squashing function $f(x) = \tanh x$ and target values $t_k = \pm 1$, a similar analysis would apply. For example, for $f(x) = \tanh(\beta x)$, the expansion

$$\tanh \beta x = \beta x - \frac{\beta^3 x^3}{3} + O(\beta^5 x^5)$$

just changes the constants in the above expansions.

Note that the second term discussed above may be written

$$(3.3.6) \qquad w - \frac{\beta^3 x^2 \eta (t - o) w^2}{16} = w - \mu w^2$$

the quadratic map of chaos theory. The same will be true of (the derivative of) the tanh squashing function. This second term, especially for larger x, larger η, larger β, will govern the further dynamical behavior of the Backpropagation weight changes. To better understand this, let us look now at the second iteration. For convenience let us denote the initial residual $\beta \eta (1 - o)/4$ by r. Then we had

$$w_1 = r + w_0 - r \left(\frac{\beta x}{2}\right)^2 w_0^2.$$

The sign of the first term is that of $(t - o)$, that of the second term is that of initial weight choice w_0, that of the third term is that of $(o - t)$. Note how taking the learning parameter η smaller allows w_0 to have more influence in setting the weight sign. For the moment then let us think of η small, β fixed, and input x near or equal to one, and consider the second iteration. We have

$$
\begin{aligned}
w_2 &= w_1 + (\Delta w)_1 = w_1 + \eta f'((w_1 x)(1 - o)) \\
&= w_1 + \frac{\eta \beta}{4}(1 - o)\left(1 - \frac{\beta^2 x^2 w_1^2}{4}\right) \\
(3.3.7) \qquad &= r + w_0 - r\left(\frac{\beta x}{2}\right)^2 w_0^2 + \frac{\eta \beta (1 - o)}{4} \\
&\quad - \frac{r \beta^2 x^2}{4}\left[\frac{\beta \eta (1 - o)}{4} + w_0 - \frac{\beta \eta (1 - o)}{4}\left(\frac{\beta x}{2}\right)^2 w_0^2\right]^2 .
\end{aligned}
$$

Let us assume that w_0 was initially small enough that we may ignore the last two terms in the bracket $[\cdot]$. Then inserting the initial residual r we have

$$(3.3.8) \qquad w_2 \cong \frac{2\eta \beta (1 - o)}{4} - \left(\frac{\beta \eta (1 - o)}{4}\right)^2 \frac{\beta^2 x^2}{4}$$

where we have also neglected the term $w_0 - r(\beta x/2)^2 w_0^2$ in the above expression, for small w_0. This last expression (3.3.8) is of interest to us for the following reason.

It is well known (although sometimes rediscovered) that the gain β and the learning parameter η play equivalent roles in sigmoid learning schemes, such as Backpropagation. This is evident from the well known property $f'(x) = \beta f(x)f(1 - x)$ of the sigmoid, from which the product $\eta\beta$ enters into the weight change expressions. However, the expression for w_2 above shows us that gain β plays an additional role, implemented by the last factor $\beta^2 x^2/4$ in the expression for w_2.

In other words, the role of gain becomes more important in the succeeding iterations of the algorithm. What we have just shown for w_2 repeats itself throughout the intermediate dynamics. For later the successive weights will be

$$(3.3.9) \qquad w_{n+1} = \frac{\beta\eta(1 - o)}{4}\left[1 - \frac{\beta^2 x^2}{4}\,w_n^2\right] + w_n$$

and the sign will not change from that of the previous w_n, if the latter has established a sign. In the in-between times, a weight sign change will depend on whether what we shall refer to as the quadratic gain term $\beta^2 x^2 w^2/4$ exceeds 1, and by how much.

Let us look a bit closer at the effective operative dynamic gain β which arises in the iterations in Backpropagation. For simplicity let us take the specified η and β parameters $\eta = \beta = 1$. Then consider the second term (3.3.6). The coefficient μ as it increases from zero to two in the quadratic map theory portends the transition from very convergent to chaotic behavior. To proceed we must now separate the input x behavior from that of the weight w behavior. Given that, our effective μ becomes

$$(3.3.10) \qquad \frac{\beta^3 x^2 \eta(t - o)}{16}$$

which is 0 or near 0 (the good, convergent behavior) only when x is 0 or near 0 (meaning that the input to this node is negligible and this node is becoming not essential to the network fit of the training input), or when the residual error $(t - o)$ has become small (meaning that we are near convergence for that node). On the other hand, our effective μ is, e.g., near 2 when $\beta^3 x^2 \eta(t - o) \approx 32$ which can occur in intermediate stages of the Backpropagation iterations when $t - o$ is still significant, specified gain β is not too small, and the net input x is relatively large. Since large gain parameters β are seldom used in neural networks, the notion of (relatively) large input x is now seen to create a large operative gain within the network.

That this prediction of logistic chaos governing neural dynamics is borne out in practice is illustrated in Fig. II-3. There the socalled exclusive-or learning problem was simulated on a neural network with two input nodes, two hidden nodes, and one output node. Including weights to the biases, there are a total of nine weights to be learned. The weight increments are plotted for 150 iterations of the backpropagation learning algorithm. As the effective dynamic gain β increases, onset of chaos is observed, with all weight changes exhibiting logistic map iterative behavior in order to quickly and adequately search weight space for an acceptable solution. It would be pure speculation to wonder if nature has built into the synapses of our brains a similar response, for example to optimize learning by adjusting effective dynamic gain, or in cases of over-input to accept probability violation with its inherent information loss.

REMARK 3.3.1. Robustness of neural learning is sometimes attributed to high network connectivity, e.g., see Hertz, Krogh, Palmer [1]. While that is a factor, the explanation here is more fundamental, and applies to even a single node. Also, while chaos in neural networks has been mentioned several times in the physics literature those instances have been for large statistical systems or socalled Hopfield nets, and do not address the basic neural learning dynamics addressed here. For more comments on the related literature, see Gustafson [9].

Comments and Bibliography

There is so much literature about the topics discussed in this Part II, certainly thousands of papers and hundreds of books, that we have made no attempt to survey it. It is hoped that the Remarks in the text provide some focus.

The paper Misra, Prigogine and Courbage [1], and its eventual transpose Antoniou and Gustafson [1], provide good entry points for Section 1.1 and Section 1.2, respectively. For Section 1.3 see Gustafson [7].

Beyond the recall of the relevant facts about wavelet theory and that of stochastic processes, most of the discussion in Section 2.1 and Section 2.2 is new, Antoniou and Gustafson [3]. Similarly, the Time Operator result of Section 2.3, Antoniou and Gustafson [2], is new and appears here for the first time.

A good survey giving also the flavor of the recent extensive interest in chaos in dynamical systems from the physics point of view, is that of Eckmann and Ruelle [1]. The new gap theory for information dimension presented in Section 3.2 is due to Gustafson [6]. The ideas for both the new neural learning dynamics theory of Section 3.3 and the quantum microirreversibility model of Section 1.3 were known at the time of Gustafson [5] but are only now being published, see Gustafson [7], [9].

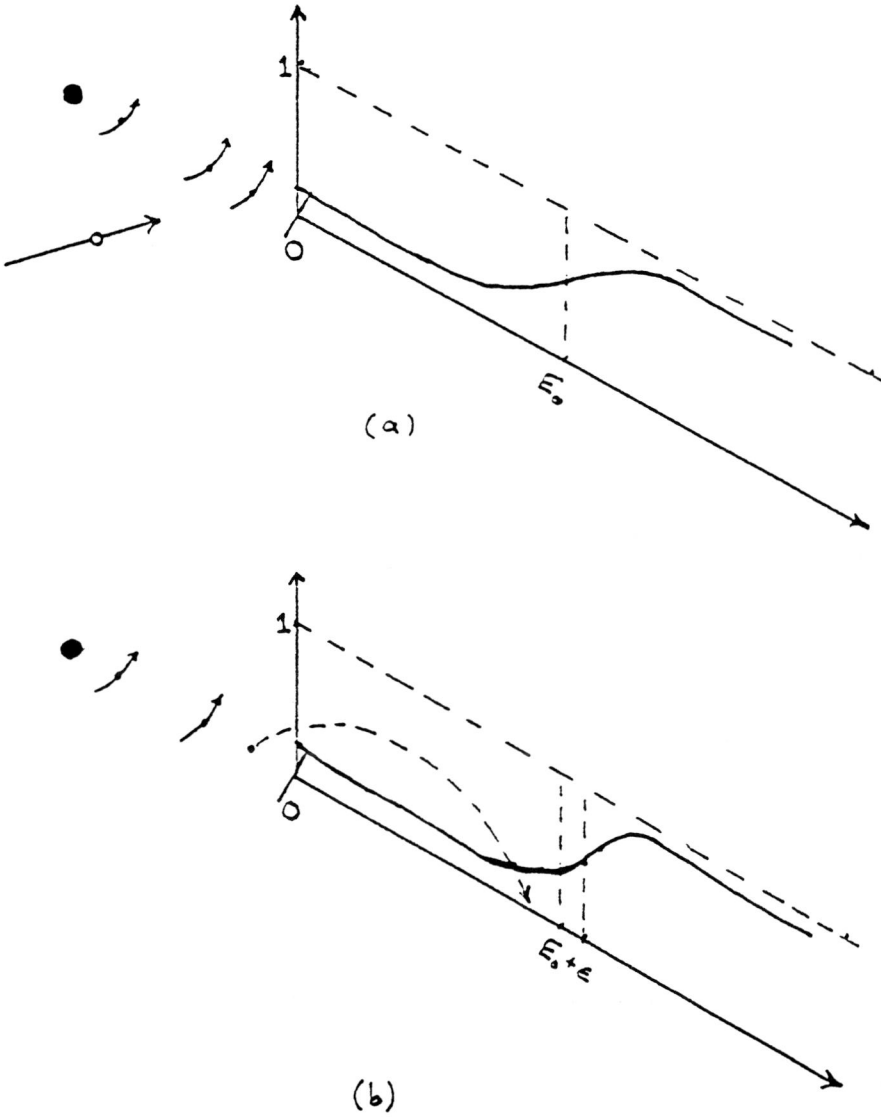

Fig. II-1. Loss of quantum probability, a microscopically irreversible event. From Gustafson [7].

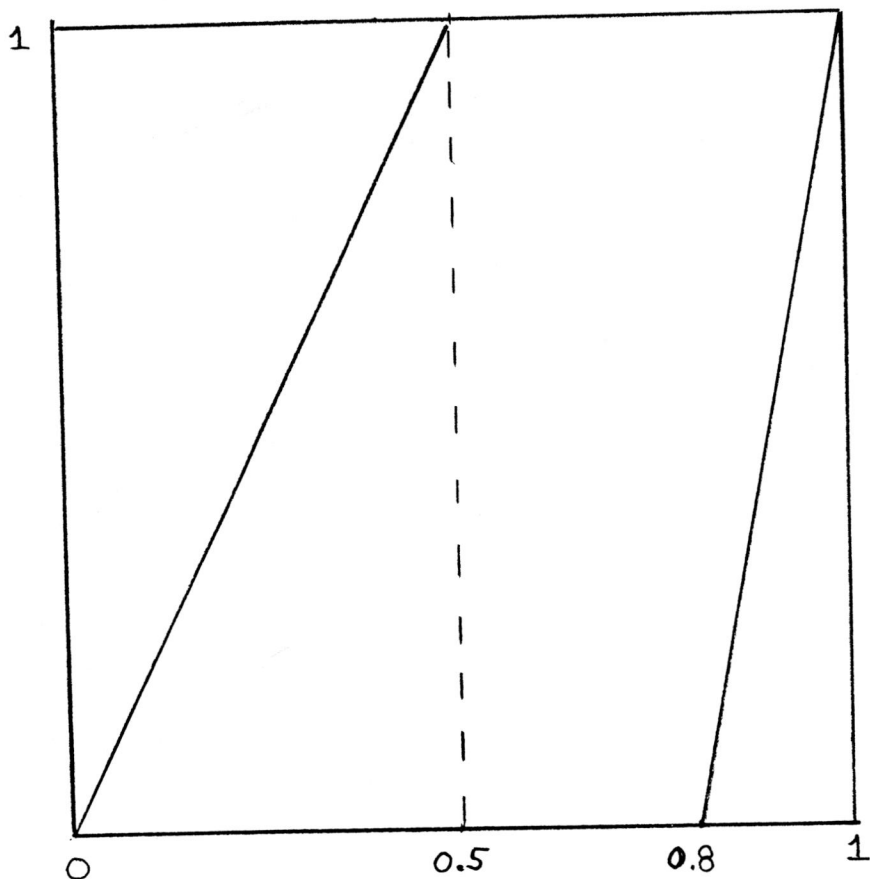

Fig. II-2. An accretive escaping map.
From Gustafson [8].

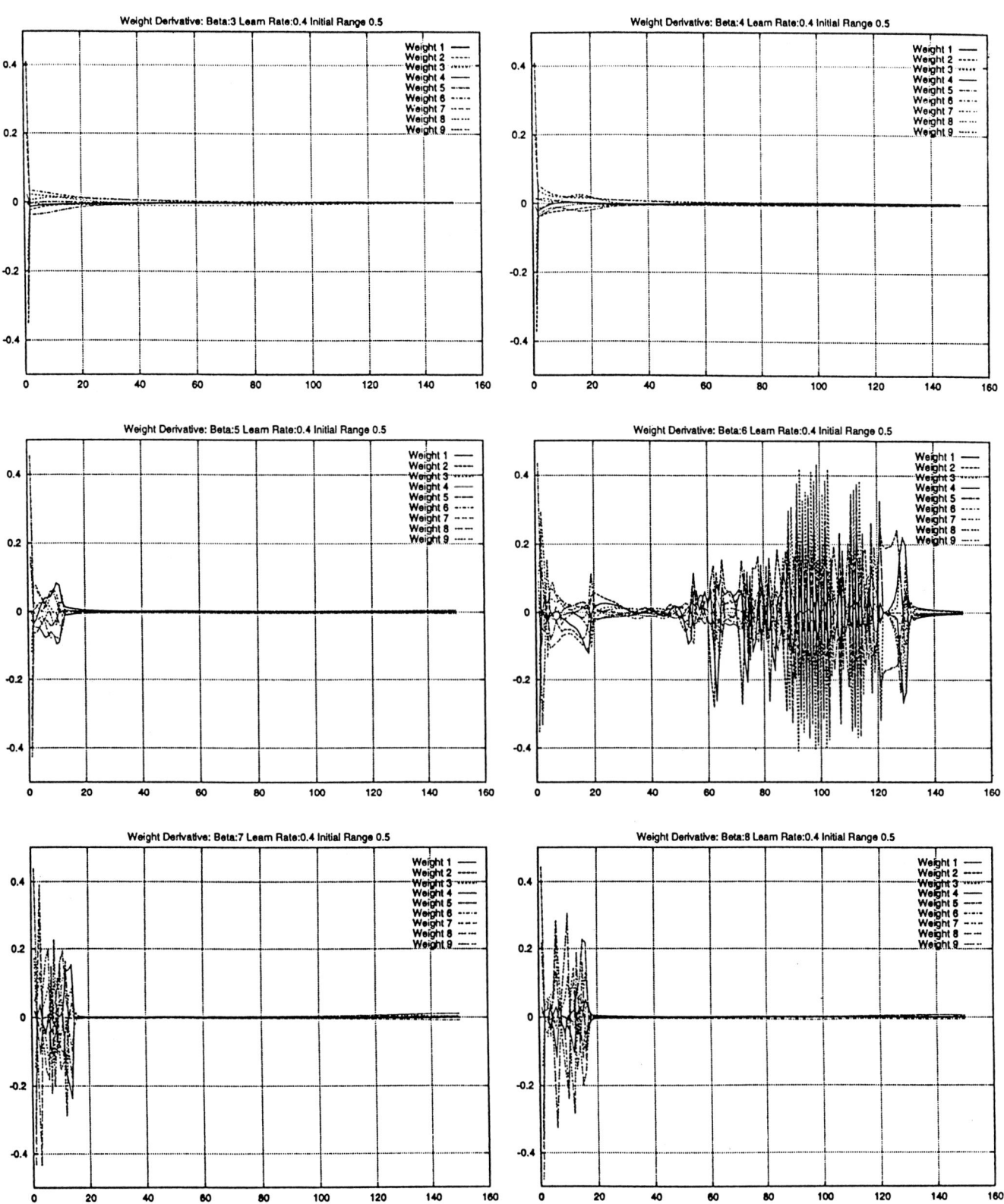

Fig. II-3. Chaos onset in neural learning.
From Gustafson [9].

112

References

E. Adelson

See P. Burt and E. Adelson.

M. Akcoglu

[1] Positive contractions on L_1-spaces, *Math. Zeit.* **143** (1975), 1–13.

M. Akcoglu and L. Sucheston

[1] On convergence of iterates of positive contractions in L_p-spaces, *J. Approximation Theory* **13** (1975), 348–362.

I. Antoniou

See S. Tasaki, Z. Suchanecki and I. Antoniou.

I. Antoniou and K. Gustafson

[1] From probabilistic descriptions to deterministic dynamics, *Physica A* **197** (1993), 153–166.

[2] The Time Operator of wavelets, to appear.

[3] Wavelets and Stochastic Processes, to appear.

[4] From Irreversible Markov Semigroups to Chaotic Dynamics, *Physica A*, to appear.

I. Antoniou, K. Gustafson, and Z. Suchanecki

[1] Dilation of stationary Markov Processes to dynamical systems, to appear.

I. Antoniou and B. Misra

[1] Relativistic internal time operator, *Intern. J. of Theor. Physics* **31** (1992), 119–136.

S. Banach

[1] *Theorie des operations lineaires*, Monografje Matematyczne, Warsaw (1932).

T. Barnwell

See M. Smith and T. Barnwell.

T. Bohr and D. Rand

[1] The entropy function for characteristic exponents, *Physica D* **25** (1987), 387–398.

R. Bowen and D. Ruelle

[1] The ergodic theory of Axiom A flows, *Inventiones Math.* **29** (1975), 181.

W. Briggs and V. Henson

[1] Wavelets and Multigrid, *SIAM J. Sci. Comput.* **14** (1993), 506–510.

P. BURT AND E. ADELSON

[1] The Laplacian Pyramid as a Compact Image Code, *IEEE Trans. Comm.* **31** (1983), 482–540.

C. CHUI

[1] *Multivariate Splines*, SIAM Publications, Philadelphia (1988).

[2] *An Introduction to Wavelets*, Academic Press, Boston (1992).

I. CORNFELD, S. FOMIN, YA SINAI

[1] *Ergodic Theory*, Springer, Berlin (1982).

M. COURBAGE

See S. Goldstein, B. Misra and M. Courbage; B. Misra, I. Prigogine and M. Courbage.

P. COVENEY AND R. HIGHFIELD

[1] *The Arrow of Time*, Fawcett Columbini, New York (1990).

H. CRAMER

[1] Stochastic processes as curves in Hilbert space, *Theor. Probability Appl.* **9** (1964), 193–204.

I. DAUBECHIES

[1] *Ten Lectures on Wavelets*, SIAM Publications, Philadelphia (1992).

[2] Orthonormal bases of compactly supported wavelets, *Comm. Pure Appl. Math.* **41** (1988), 909–996.

[3] Orthonormal bases of wavelets with finite support-connection with discrete filters, in *Wavelets* (J. Combes, A. Grossmann, P. Tchamitchian, eds.), Springer, Berlin (1989), 38–66.

E. B. DAVIES

[1] *Quantum Theory of Open Systems*, Academic, London (1976).

[2] *One Parameter Semigroups*, Academic, London (1980).

G. DE RHAM

[1] Un peu de mathematiques a propos d'une courbes plane, *Elemente der Mathematik* **2** (1947), 73–76.

[2] Sur quelques courbes definies par des equations fonctionnelles, *Rend. Sem. Mat. Torino* **16** (1957), 101–112.

114

R. Devaney

[1] An Introduction to Chaotic Dynamical Systems, Benjamin–Cummings, Menlo Park (1986).

J. Doob

[1] *Stochastic Processes*, Wiley, New York (1953).

D. Driebe

See H. Hasegawa and D. Driebe.

J. P. Eckmann and D. Ruelle

[1] Ergodic theory of chaos and strange attractors, *Reviews of Modern Physics* **57** (1985), 617–656.

K. Falconer

[1] *Fractal Geometry*, Wiley, New York (1990).

R. Feynman, R. Leighton and M. Sands

[1] *The Feynman Lectures on Physics*, Vol. I, Addison–Wesley, Reading, Mass. (1989).

C. Foias

See B. Sz. Nagy and C. Foias.

S. Fomin

See I. Cornfeld, S. Fomin, Ya Sinai.

P. Frederickson, J. Kaplan, E. Yorke and J. Yorke

[1] The Lyapunov dimension of strange attractors, *J. Diff. Eqns.* **49** (1983), 185–207.

D. Gabor

[1] Theory of communication, *J. IEE* **93** (1946), 429–457.

S. Goldstein, B. Misra and M. Courbage

[1] On intrinsic randomness of dynamical systems, *J. Stat. Phys.* **25** (1981), 111–126.

R. Goodrich

See K. Gustafson and R. Goodrich; K. Gustafson, R. Goodrich and B. Misra.

R. Goodrich, K. Gustafson and B. Misra

[1] A converse to Koopman Lemma, *Physica* **102A** (1980), 379–388.

[2] On K-Flows and irreversibility, *J. Stat. Phys.* **43** (1986), 317–320.

P. Grassberger

See H. Kantz and P. Grassberger.

K. GUSTAFSON

[1] The counter problem, (1975), unpublished.

[2] Irreversibility questions in chemistry, quantum-counting, and time-delay, in *Energy Storage and Redistribution in Molecules* (J. Hinze, ed.), Plenum, New York (1983), 531–541.

[3] *Partial Differential Equations*, Wiley, New York (1980, 1987), Kaigai, Tokyo (1991, 1992), International Journal Services, Calcutta (1993).

[4] Four principles of vortex motion, in *Vortex Methods and Vortex Motion* (K. Gustafson, J. Sethian, eds.), SIAM, Philadelphia (1990), 95–141.

[5] Reversibility in Neural Processing Systems, in *Statistical Mechanics of Neural Networks* (L. Garrido, ed.), Lecture Notes in Physics **368**, Springer, Berlin (1990), 269–285.

[6] A note on optimal ergodic invariant measures for piecewise linear maps on fractal repellers, *Physics Letters A* **208** (1995), 117–126.

[7] Operator spectral states, *Computers and Mathematics with Applications* (1996), to appear.

[8] A Gap Theory of Information Dimension for Fractal Accretive Maps, to appear.

[9] Robustness of the Backpropagation Neural Net Learning Algorithm, *Mathematical Modelling and Scientific Computing* **7** (1996), to appear.

See also I. Antoniou and K. Gustafson; R. Goodrich, K. Gustafson and B. Misra.

K. GUSTAFSON AND R. GOODRICH

[1] A Banach–Lamperti theorem and similarity transformations in statistical mechanics, *Colloq. Math. Soc. Janos Bolyai* **35** (1980), 567–579.

[2] Kolmogorov systems and Haar systems, *Colloq. Math. Soc. Janos Bolyai* **49** (1985), 401–416.

K. GUSTAFSON, R. GOODRICH AND B. MISRA

[1] Irreversibility and stochasticity of chemical processes, in *Quantum Mechanics in Mathematics, Chemistry, and Physics* (K. Gustafson and W. Reinhardt, eds.), Plenum, New York (1991), 203–210.

K. GUSTAFSON AND B. MISRA

[1] Canonical commutation relations of quantum mechanics and stochastic regularity, *Letters in Math. Phys.* **1** (1976), 275–280.

116

K. GUSTAFSON AND D. RAO

[1] *Numerical Range: The Field of Values of Operators and Matrices*, to appear.

A. HAAR

[1] Zur Theorie der Orthogonalen Functionensystem I, II, *Math. Annalen* **69** (1910), 38–53; **71** (1911), 331–371.

P. HALMOS

[1] *A Hilbert Space Problem Book*, 2nd Ed., Springer, New York (1982).

H. HASEGAWA AND D. DRIEBE

[1] Intrinsic irreversibility and the validity of the kinetic description of chaotic systems, *Physical Review E* **50** (1994), 1781–1809.

S. HAWKING

[1] *A Brief History of Time*, Bantam, New York (1988).

C. HEIL AND D. WALNUT

[1] Continuous and discrete wavelet transforms, *SIAM Review* **31** (1989), 628–666.

V. HENSON

See W. Briggs and V. Henson.

J. HERTZ, A. KROGH AND R. PALMER

[1] *Introduction to the Theory of Neural Computation*, Addison–Wesley, Reading, MA (1991).

T. HIDA

[1] *Stationary Stochastic Processes*, Princeton Press, Princeton, New Jersey (1970).

R. HIGHFIELD

See P. Coveney and R. Highfield.

C. IONESCU-TULCEA

[1] Mesures dans les espaces produits, *Atti Accad. Naz. Lincei Rend. Cl. Sci. Fis. Mat. Nat.* (1949), 208–211.

B. JAWERTH AND W. SWELDENS

[1] An overview of wavelet based multiresolution analyses, *SIAM Review* **36** (1994), 377–412.

L. KADANOFF AND C. TANG

[1] Escape from strange repellers, *Proc. Nat. Acad. Sci. USA* **81** (1984), 1276.

S. KAKUTANI

See K. Yosida and S. Kakutani.

H. KANTZ AND P. GRASSBERGER

[1] Repellers, semi-attractors, and long lived chaotic transients, *Physica* **17D** (1985), 75–86.

J. KAPLAN

See P. Frederickson, J. Kaplan, E. Yorke and J. Yorke.

B. KOOPMAN

[1] Hamiltonian systems and transformations in Hilbert spaces, *Proc. Nat. Acad. Sci. USA* **17** (1931), 315–318.

B. KOOPMAN AND J. VON NEUMANN

[1] Dynamical systems of continuous spectra, *Proc. Nat. Acad. Sci. USA* **18** (1932), 255–266.

P. KOPP

[1] *Martingales and Stochastic Integrals*, Cambridge Press, Cambridge, London (1984).

U. KRENGEL

[1] *Ergodic Theorems*, De Gruyter, Berlin (1985).

A. KROGH

See J. Hertz, A. Krogh and P. Palmer.

J. LAMPERTI

[1] On the isometries of certain function spaces, *Pac. J. Math.* **8** (1958), 459–466.

A. LASOTA AND M. MACKEY

[1] *Probabilistic Properties of Deterministic Systems*, Cambridge Press, Cambridge (1985).

P. LAX AND R. PHILLIPS

[1] *Scattering Theory*, Academic, New York (1967).

R. LEIGHTON

See R. Feynman, R. Leighton and M. Sands.

J. LEWIS AND L. THOMAS

[1] How to make a heat bath, in *Functional Integration* (A. Arthurs, ed.), Clarendon, Oxford (1975), 97–123.

G LUDWIG

[1] *Foundations of Quantum Mechanics* I, II, Springer, Berlin (1983).

G. MACKEY

118

[1] *Theory of Group Representations*, U. of Chicago Press, Chicago (1976).

M. MACKEY

[1] *Times Arrow: The Origins of Thermodynamic Behavior*, Springer, New York (1992).

See also A. Lasota and M. Mackey.

S. MALLAT

[1] Multiresolution approximations and wavelet orthonormal bases of $L^2(R)$, *Trans. Amer. Math. Soc.* **315** (1989), 69–87.

B. MANDELBROT

[1] *The Fractal Geometry of Nature*, Freeman, New York (1983).

P. MATHEWS

See E. Sudershan, P. Mathews and J. Rao.

R. MAY

[1] Simple Mathematical models with very complicated dynamics, *Nature* **261** (1976), 459–467.

Y. MEYER

[1] *Wavelets and Operators*, Cambridge Press, Cambridge (1992).

B. MISRA

[1] Nonequilibrium entropy, Lyapunov variables, and ergodic properties of classical systems, *Proc. Nat. Acad. Sci. USA* **75** (1978), 1627–1631.

See also I. Antoniou and B. Misra; S. Goldstein, B. Misra and M. Courbage; R. Goodrich, K. Gustafson and B. Misra; K. Gustafson and B. Misra.

B. MISRA AND I. PRIGOGINE

[1] On the foundations of kinetic theory, *Supplement, Progress Theor. Phys.* **69** (1980), 101–110.

[2] Time, probability, and dynamics, in *Long Time Prediction in Dynamics* (C. Horton, L. Reichl, A. Szebehely, eds.), Wiley, New York (1983), 21–43.

B. MISRA, I. PRIGOGINE AND M. COURBAGE

[1] From deterministic dynamics to probabilistic descriptions, *Physica* **98A** (1979), 1–26.

[2] Lyapounov variable: entropy and measurement in quantum mechanics, *Proc. Natl. Acad. Sci. USA* **76** (1979), 4768–4772.

B. MISRA AND E. SUDERSHAN

[1] The Zeno's paradox in quantum theory, *J. Math. Phys.* **18** (1977), 756–763.

H. MORI, B. SO, T. OSE

[1] Time correlation functions of one-dimensional transformations, *Progress Theor. Phys.* **66** (1981), 1266–1283.

J. NEVEU

[1] Relations entre la theorie des martingales et la theorie ergodic, *Ann. Inst. Fourier Grenoble* **15** (1965), 31–42.

G. NICOLIS AND I. PRIGOGINE

[1] *Exploring Complexity*, Freeman, New York (1989).

H. NUSSENZVEIG

[1] *Causality and Dispersion Relations*, Academic, New York (1972).

T. OSE

See H. Mori, B. So, and T. Ose.

R. PALMER

See J. Hertz, A. Krogh and R. Palmer.

W. PAULI

[1] Die allgemeinen Prinzipien der Wellenmechanik, in *Handbuch der Physik* (S. Flugge, ed.), Springer, Berlin (1958), 1–168.

R. PENROSE

[1] *The Emperor's New Mind*, Oxford Press, Oxford (1989).

R. PHILLIPS

See P. Lax and R. Phillips

I. PRIGOGINE

See B. Misra and I. Prigogine; B. Misra, I. Prigogine and M. Courbage; G. Nicolis and I. Prigogine.

I. PRIGOGINE AND P. RESIBOIS

[1] On the kinetics of the approach to equilibrium, *Physica* **27** (1961), 629–646.

I. PRIGOGINE AND I. STENGERS

[1] *Order Out of Chaos*, Bantam, Toronto (1984).

H. PRIMAS

[1] *Chemistry, Quantum Mechanics and Reductionism*, Springer Lecture Notes in Chemistry **24**, Berlin (1981).

C. PUTNAM

[1] *Commutation Properties of Hilbert Space Operators and Related Topics*, Springer, Berlin (1967).

D. RAND

See T. Bohr and D. Rand.

D. RAO

See K. Gustafson and D. Rao.

J. RAO

See E. Sudershan, P. Mathews and J. Rao.

M. RAO

[1] Abstract Martingales and Ergodic Theory, in *Multivariate Analysis III*, Academic Press, New York (1972), 45–60.

P. RESIBOIS

See I. Prigogine and P. Resibois.

E. ROBINSON

[1] Predictive decomposition of time series with application to seismic exploration, in *Deconvolution* (G. Webster, ed.), Society of Exploration Geophysicists, Tulsa, Oklahoma (1978), 52–118.

V. ROKHLIN

[1] Exact endomorphisms of Lebesgue spaces, *Amer. Math. Soc. Transl.* **39** (1964), 1–36.

D. RUELLE

[1] A measure associated with Axiom A attractors, *Amer. J. Math.* **98** (1976), 619.

See also R. Bowen and D. Ruelle; J. Eckmann and D. Ruelle.

M. SANDS

See R. Feynman, R. Leighton and M. Sands.

L. SCHIFF

[1] *Quantum Mechanics*, McGraw Hill, New York (1968).

Y. SINAI

[1] Gibbs measures in ergodic theory, *Usp. Mat. Nauk.* **27** (1972), 21.

See also I. Cornfeld, S. Fomin and Y. Sinai.

M. SMITH AND T. BARNWELL

[1] Exact reconstruction techniques for tree-structured subband coders, *IEEE Trans. Acoustics* **34** (1986), 434–441.

B. SO

See H. Mori, B. So and T. Ose.

I. STENGERS

See I. Prigogine and I. Stengers.

G. STRANG

[1] Wavelets and dilation equations: a brief introduction, *SIAM Review* **31** (1989), 614–627.

[2] Wavelets, *American Scientist* **82** (1994), 250–255.

Z. SUCHANECKI

See I. Antoniou, K. Gustafson and Z. Suchanecki; S. Tasaki, Z. Suchanecki, and I. Antoniou.

L. SUCHESTON

See M. Akcoglu and L. Sucheston.

E. SUDERSHAN

See B. Misra and E. Sudershan.

E. SUDERSHAN, P. MATHEWS AND J. RAO

[1] Stochastic dynamics of quantum mechanical systems, *Phys. Rev.* **121** (1961), 920–924.

See also B. Misra and E. Sudershan.

W. SWELDENS

See B. Jawerth and W. Sweldens.

B. SZ. NAGY AND C. FOIAS

[1] *Harmonic Analysis of Operators on Hilbert Space*, North Holland, Amsterdam (1970).

C. TANG

See L. Kadanoff and C. Tang.

S. TASAKI, Z. SUCHANECKI AND I. ANTONIOU

122

[1] Ergodic properties of piecewise linear maps on fractal repellers, *Physics Letters A* **179** (1993), 103–110.

L. THOMAS

See J. Lewis and L. Thomas.

S. ULAM AND J. VON NEUMANN

[1] On combination of stochastic and deterministic processes, *Bull. Amer. Math. Soc.* **53** (1947), 1120.

J. VON NEUMANN

See B. Koopman and J. Von Neumann; S. Ulam and J. Von Neumann.

D. WALNUT

See C. Heil and D. Walnut.

P. WALTERS

[1] *Ergodic Theory*, Springer Lecture Notes in Mathematics #458, Berlin (1975).

E. YORKE

See P. Frederickson, J. Kaplan, E. Yorke and J. Yorke.

J. YORKE

See P. Frederickson, J. Kaplan, E. Yorke and J. Yorke.

K. YOSIDA AND S. KAKUTANI

[1] Operator-theoretical treatment of Markoff process and mean ergodic theorems, *Ann. of Math.* **42** (1941), 188–228.

PART III
Recent Developments in Linear Algebra

Perspective. In this third Part, we present the essentials of a new part of Linear Algebra, a theory of antieigenvalues. Antieigenvalues measure the principal turning angles of linear operators or matrices, just as eigenvalues measure the principal dilation properties of operators or matrices. Although the basic concepts about antieigenvalues are now known, much remains to be learned beyond the case in which the operator or matrix is selfadjoint or normal.

Chapter 1 presents the original (1966–1969) motivations in operator semigroup theory, which led to the concept of antieigenvalue. In that setting the first antieigenvalue was called $\cos A$. The early developments centered on semigroup theory and to a lesser extent on creating an operator trigonometry. A key early (1968) finding was the minmax theorem, through which $\sin A$ may be meaningfully defined. This minmax theorem has been recently (1995) revisited and provides a new way to approximately construct antieigenvectors for general operators.

Chapter 2 brings forth antieigenvalues and their corresponding antieigenvectors in their own right. A main result is the Euler equation for the antieigenvectors. This may be viewed as a significant extension of the Rayleigh–Ritz Theorem for the variational characterization of eigenvectors. This Euler equation was known and mentioned in 1969 but was not published in the open literature until twenty-five years later, in 1994. Chapter 2 also contains what we believe to be the first exposition of the true geometrical meaning of the Kantorovich–Wielandt inequalities

Chapter 3 exposes and explores the recently found connections between the operator trigonometry and several iterative methods for solving $Ax = b$. The latter are becoming much more important as physical modelling moves from two to three space dimensions and the resulting linear systems become very large. Chapter 3 also contains numerical evidence for a combinatorial theory of higher antieigenvalues and antieigenvectors. A possible use of the first antieigenvalue μ for designing preconditioners for linear systems is also discussed.

Chapter 1   Operator Trigonometry

1.1. Semigroup Generators. The operator trigonometry presented here originated in 1966–1969 in Gustafson [1], [2], [3], [4], [5], [6], [7] in connection with perturbation questions in the Hille–Yosida theory of operator semigroups, see Yosida [1] or Kato [1]. Let $G(1,0)$ denote the set of infinitesimal generators A of contraction semigroups $T(t)$ on a Hilbert or Banach space X. The following characterization of $G(1,0)$ is known.

THEOREM 1.1.1 (Lumer–Phillips). Let A be densely defined with domain $D(A)$ and range $R(A)$ in X. Then A generates a continuous contraction semigroup on X if and only if

$$(1.1.1) \qquad \operatorname{Re}[Ax, x] \leqq 0 \qquad x \in D(A)$$

$$(1.1.2) \qquad R(I - A) = X$$

Proof. See Lumer and Phillips [1], or Yosida [1]. Theorem 1.1.1 is a useful version of the general Hille Yosida Theorem characterizing more general semigroup generators.

Condition (1.1.1) is called: A is *dissipative*. There the bracket $[y, x]$ denotes any semi-inner product defined on X. Of course if X is a Hilbert space, there is only one semi-inner product compatible with the norm, namely the usual inner product $\langle y, x \rangle$. For simplicity, in the rest of this Part III, we shall always assume that we are in a Hilbert space X.

It should be remarked that the operator trigonometry is not much developed in the more general Banach Space or Topological Vector Space settings, even though all essential trigonometric entities may be immediately defined there. Hopefully such extended theory will be motivated by applications in such more general settings.

A well known important perturbation theorem for contraction semigroup generators is the following.

THEOREM 1.1.2 (Kato–Rellich). If $A \in G(1,0)$ and B is a relatively small perturbation, namely

$$(1.1.3) \qquad \|Bx\| \leqq a\|x\| + b\|Ax\|, \qquad x \in D(A),\ b < 1$$

then the sum $A + B \in G(1,0)$ if and only if $A + B$ is dissipative.

Proof. See Kato [1] and Gustafson [1], [8]. Through this theorem one may, for example, establish the selfadjointness of the basic Schrödinger partial differential operators of quantum mechanics.

A lesser known multiplicative perturbation theorem for contraction semigroup generators is the following.

THEOREM 1.1.3 (Dorroh–Gustafson). If $A \in G(1,0)$ and B is a bounded operator satisfying

$$\|\epsilon B - I\| < 1 \qquad \text{some} \quad \varepsilon > 0 \tag{1.1.4}$$

then the product $BA \in G(1,0)$ if and only if BA is dissipative.

Proof. See Gustafson [4]. But the proof follows rather directly from Theorem 1.1.2 by writing

$$\epsilon BA = A + (\epsilon B - I)A \tag{1.1.5}$$

and regarding the second term as an additive perturbation.

The condition (1.1.4) is easily seen to be equivalent to

$$\text{Re} \langle Bx, x \rangle \geqq m > 0 \tag{1.1.6}$$

for all x such that $\|x\| = 1$. Condition (1.1.6) is called: B is *strongly accretive*. Once this is seen, Theorem 1.1.3 takes on a particularly appealing meaning: A must be negative (i.e., dissipative), B must be positive (i.e., accretive), and then, when is BA negative (i.e., dissipative)?

Let us consider this question by looking more closely at (1.1.5). We may write

$$\begin{aligned}
\text{Re} \langle \epsilon BAx, x \rangle &= \text{Re} \langle (\epsilon B - I)Ax, x \rangle + \text{Re} \langle Ax, x \rangle \\
&\leqq \|\epsilon B - I\| \|Ax\| \|x\| + \text{Re} \langle Ax, x \rangle
\end{aligned} \tag{1.1.7}$$

from which we have

THEOREM 1.1.4. Given B bounded strongly accretive and A dissipative, then BA is dissipative if

$$\min_{\epsilon > 0} \|\epsilon B - I\| \leqq \inf_{x \in D(A)} \frac{\text{Re} \langle (-A)x, x \rangle}{\|Ax\| \|x\|} \tag{1.1.8}$$

Proof. Transpose and divide from (1.1.7).

It was from (1.1.8) that the operator trigonometry ensued. This will be described in the next section.

REMARK 1.1.5. Perturbation of more general semigroups, e.g., of type $G(M, \beta)$, may be considered in a similar way. Usually this is done by translation in the complex plane to the cases $G(1,0)$ or $G(1, \beta)$, if possible.

1.2. Cos A and Sin A. Let A be a strongly accretive operator, e.g., A is the $(-A)$ of the expression (1.1.8).

DEFINITION:

$$(1.2.1) \qquad \cos A = \inf_{x \in D(A)} \frac{\operatorname{Re} \langle Ax, x \rangle}{\|Ax\| \|x\|}$$

Let B be a strongly accretive bounded operator.

DEFINITION:

$$(1.2.2) \qquad \sin B = \inf_{\epsilon > 0} \|\epsilon B - I\|$$

Using these entities we may state a very sharp criteria for the product of two positive (accretive) operators to be accretive.

THEOREM 1.2.1. Let A and B be two bounded strongly accretive operators or matrices on a Hilbert space X. Then BA is accretive if

$$(1.2.3) \qquad \sin B \leqq \cos A$$

Proof. Apply Theorem 1.1.4.

The definition $\cos A$ is very natural from Schwarz's Inequality, although it required the semigroup perturbation questions of Section 1.1 to induce it. It should be noted that $\cos A$ is a completely geometrical notion and is not the entity $\cos(A)$ defined as a power series in a functional calculus of A. Through $\cos A$ we have

DEFINITION: The angle $\phi(A)$ of an operator A is the angle in the first quadrant whose cosine is $\cos A$.

In other words, $\phi(A)$ is the supremum of all possible angles through which A may turn a vector x, over all x in $D(A)$. We will return to this interpretation at several points later in this chapter.

It is not obvious that $\sin B$ as defined in (1.2.2) makes sense as a trigonometric entity. This will be established in the next section, by showing that $\sin B = \sqrt{1 - \cos^2 B}$.

Expressions for $\cos A$ and $\sin B$ are known for positive definite selfadjoint operators A and B, namely

$$(1.2.4) \qquad \cos A = \frac{2\sqrt{Mm}}{M + m}, \quad \sin B = \frac{M - m}{M + m}$$

where M and m are the upper and lower bounds of such operators, viz., $M = \lambda_{\max} =$ the largest eigenvalue, and $m = \lambda_{\min} =$ the lowest eigenvalue, in the case of matrices. Their derivation will be discussed in Chapter 2.

The sharpness of Theorem 1.2.1 is evidenced by the following example.

EXAMPLE. 1.2.2. Let A and B be symmetric positive definite matrices, with $M_A = 1$, $m_A = 1/2$, and $M_B = 1$. Then BA is accretive provided that $m_B \geq 0.0295$.

Proof. Apply (1.2.3) of Theorem 1.2.1, and (1.2.4).

Theorem 1.2.1 thus indicates that 'most' products of positive operators are accretive, even when they do not commute.

REMARK 1.2.3. It must be admitted that explicit expressions for $\cos A$ and $\sin A$ such as those in (1.2.4) are lacking for general operators A.

REMARK 1.2.4. Given a cosine and a sine, one has a trigonometry, in the sense that the other four trigonometric functions may be defined in terms of the two primitive ones. However, a little experimentation shows that most of the common trigonometric identities fail for this operator trigonometry. We conjecture, Gustafson [16], that the operator trigonometry forms a kind of spherical hyperbolic geometry.

1.3. The Min-Max Theorem. As mentioned above, the key to the operator trigonometry was the discovery of the min-max theorem, through which $\sin B$ made sense.

THEOREM 1.3.1. For any strongly accretive bounded operator B on a Hilbert space X,

$$(1.3.1) \qquad \sup_{\|x\|=1} \inf_{-\infty < \epsilon < \infty} \|(\epsilon B - I)x\|^2 = \inf_{-\infty < \epsilon < \infty} \sup_{\|x\|=1} \|(\epsilon B - I)x\|^2$$

In particular

$$(1.3.2) \qquad \sin B = \inf_{\epsilon > 0} \|\epsilon B - I\|$$

where $\sin B$ is defined as $\sin B = \sqrt{1 - \cos^2 B}$.

Proof. Let us introduce the notation

$$(1.3.3) \qquad g_m(B) = \min_{\epsilon > 0} \|\epsilon B - I\|$$

for the right hand side of (1.3.2). The fact the infimum is obtained follows by elementary compactness arguments.

To prove Theorem 1.3.1, let us first note that the right hand side of (1.3.1) is just $g_m^2(B)$. We will show below that this minimum is attained uniquely. Also we note that the left hand side of (1.3.1) is indeed $1 - \cos^2 B$. To see this, consider the parabola

$$(1.3.4) \qquad \|(\epsilon B - I)x\|^2 = \epsilon^2 \|Bx\|^2 - 2\epsilon \mathrm{Re}\,\langle Bx, x\rangle + 1$$

This parabola in ϵ achieves its minimum at $\epsilon_m(x) = \mathrm{Re}\,\langle Bx, x\rangle / \|Bx\|^2$ and the value of the minimum is

$$(1.3.5) \qquad 1 - (\mathrm{Re}\,\langle Bx, x\rangle / \|Bx\|)^2$$

The supremum over x, $\|x\| = 1$, of this quantity is $1 - \cos^2 B$.

Next let us assure ourselves that the minimum $g_m(B)$ is attained uniquely. This is the case for uniformly convex Banach spaces but not generally true otherwise. Here we give a Hilbert space proof. Let us suppose the contrary, that $\|\epsilon B - I\|$ dips down and then has a flat interval minimum. Let $\epsilon_1 < \epsilon_2$ be the ends of this interval. Then for every small $\delta > 0$ there is an x, $\|x\| = 1$, such that the parabola $\|(\epsilon B - I)x\|^2$ is less than or equal to $g_m^2(B)$ at both ϵ_1 and ϵ_2 but at their midpoint must be within δ of the minimum flat:

$$(1.3.6) \qquad \|(((\epsilon_1 + \epsilon_2)/2)B - I)x\|^2 \geqq g_m^2(B) - \delta$$

But no quadratic function of ϵ, anchored at the value 1 at $\epsilon = 0$, can satisfy this condition for arbitrarily small δ.

Let ϵ_m denote the unique $\epsilon > 0$ at which $g_m(B)$ is attained. The convex curve $\|\epsilon B - I\|^2$ is continuous in ϵ, has left and right hand derivatives for all ϵ, and these are equal except at a countable set of ϵ. Thus we may speak freely of the "slope" of $\|\epsilon B - I\|^2$ for a dense set of $\epsilon > 0$.

Consider now the curve $\|\epsilon B - I\|^2$ near its unique minimum value $g_m^2(B)$. For any chosen fixed ϵ just, but strictly, to the left of ϵ_m, $\|\epsilon B - I\|^2$ slopes downward and is strictly greater than $\|\epsilon_m B - I\|^2$. Since $\|\epsilon B - I\|^2$ is a supremum, there thus exists an x_1, $\|x_1\| = 1$, such that $\|(\epsilon B - I)x_1\|^2 > \|\epsilon_m B - I\|^2$. Moreover, we may choose this x_1 so that its minimum $\epsilon_m(x_1)$ lies (possibly nonstrictly) to the right of ϵ_m, for otherwise all of the parabolas $\|(\epsilon B - I)x_1\|^2$ increasing to achieve the supremum $\|(\epsilon B - I)\|^2$ from below would turn upward from points $\epsilon_m(x_1)$ strictly to the left of ϵ_m, cutting and thereby violating the downward sloping curve $\|\epsilon B - I\|^2$.

Similarly, there exists an x_2, $\|x_2\| = 1$, for any chosen fixed ϵ just, but strictly, to the right of ϵ_m, such that $\|(\epsilon B - I)x_2\|^2 > \|\epsilon_m B - I\|^2$ and such that $\epsilon_m(x_2)$ lies (possibly nonstrictly) to the left of ϵ_m. Moreover, we can choose x_1 and x_2 so that $\|(\epsilon B - I)x_1\|^2 \geqq g_m^2(B) - \delta$ and $\|(\epsilon_m B - I)x_2\|^2 \geqq g_m^2(B) - \delta$ for any prespecified small $\delta > 0$, so that those two parabolas pass arbitrarily close below the minimum point $g_m^2(B)$.

Consider first the case in which, for given small $\delta > 0$, $\epsilon_m(x_1)$ and $\epsilon_m(x_2)$ lie strictly to the right and left of ϵ_m, respectively. Let $x = \xi x_1 + \eta x_2$, where ξ and η are real and satisfy

$$(1.3.7) \qquad 1 = \|x\|^2 = \xi^2 + \eta^2 + 2\eta\xi \mathrm{Re}\langle x_1, x_2 \rangle.$$

Using (1.3.7), a simple computation shows that $\|(\epsilon_m B - I)x\|^2 \geq g_m^2(B) - \delta + 2\xi\eta C$, where $C = \mathrm{Re}\{\langle(\epsilon_m B - I)x_1, (\epsilon_m B - I)x_2\rangle - (g_m^2 - \delta)\langle x_1, x_2 \rangle\}$, and by restricting ξ and η in (1.3.7) to the appropriate quadrant we can assure that $2\xi\eta C \geq 0$. By choosing ξ and η not only of appropriate sign but also near one or the other coordinate axes, we can also assure at this point in the construction that this term $2\xi\eta C$ is also arbitrarily small, but it turns out better to let this happen automatically as a consequence of later steps in the construction.

Next, we would like to show that $\epsilon_m(x)$ can be made arbitrarily close to ϵ_m. Consider first the case in which we ask for exact equality $\epsilon_m(x) = \epsilon_m$. By a short computation this can be seen to be equivalent to

$$(1.3.8) \qquad \begin{aligned} &\xi^2\left\{\mathrm{Re}\langle Bx_1, x_1\rangle\left(1 - \frac{\epsilon_m}{\epsilon_1}\right)\right\} + \eta^2\left\{\mathrm{Re}\langle Bx_2, x_2\rangle\left(1 - \frac{\epsilon_m}{\epsilon_2}\right)\right\} \\ &- 2\xi\eta\mathrm{Re}\langle Bx_1, (\epsilon_m B - I)x_2\rangle = 0, \end{aligned}$$

where ϵ_1 denotes $\epsilon_m(x_1)$ and ϵ_2 denotes $\epsilon_m(x_2)$. Since $(1 - \epsilon_m/\epsilon_1)(1 - \epsilon_m/\epsilon_2) < 0$, the degenerate hyperbola (1.3.8) and the ellipse (1.3.7) have a point in common. This assures that $\epsilon_m(x) = \epsilon_m$ and that $\|(\epsilon_m B - I)x\|^2 \geqq g_m^2(B) - \delta + 2\xi\eta C$ for arbitrarily small $\delta > 0$. But since $\epsilon_m(x) = \epsilon_m$, the term $2\xi\eta C$ must be small, for otherwise $\|(\epsilon_m B - I)x\|^2$ would exceed $\|\epsilon_m B - I\|^2$, which it cannot. Thus these $x = x(\delta)$ provide the supremum sequence for the left side of (1.3.1) to equal the right side, namely, $g_m^2(B)$.

For those cases in which we cannot ask for $\epsilon_m(x)$ to be exactly ϵ_m in the above construction, but only arbitrarily close to ϵ_m, we proceed in the same way. It is desired that

$$(1.3.9) \qquad \epsilon_m(x) = \frac{\xi^2\mathrm{Re}\langle Bx_1, x_1\rangle + \eta^2\mathrm{Re}\langle Bx_2, x_2\rangle + 2\xi\eta\mathrm{Re}\langle Bx_1, x_2\rangle}{\xi^2\|Bx_1\|^2 + \eta^2\|Bx_2\|^2 + 2\xi\eta\mathrm{Re}\langle Bx_1, Bx_2\rangle} = \tilde{\epsilon}_m$$

where $\tilde{\epsilon}_m$ is to be arbitrarily close to ϵ_m. Multiplying this out results in the same expression (1.3.8) with ϵ_m replaced by $\tilde{\epsilon}_m$. Provided that we are not in an exceptional instance in which $\epsilon_1 = \epsilon_2$, we may ask for $\tilde{\epsilon}_m$ arbitrarily close to ϵ_m strictly between ϵ_2 and ϵ_1, and proceed as before.

The special cases in which $\epsilon_m(x_1)$ or $\epsilon_m(x_2)$ coincide with ϵ_m correspond to one or both branches of (1.3.8) lying along ξ or η corodinate axes. In that case one chooses $\xi = 0$ or $\eta = 0$ as the case may be; it happens then that the $2\xi\eta C$ term is killed exactly. The special case in which $x_1 = x_2$ corresponds to a degenerate parabola in (1.3.7). These two parallel lines span all four quadrants so the sign choice on ξ, η remains unrestricted. Thus in all cases we have constructed x with $\epsilon_m(x)$ arbitrarily close to ϵ_m and $\|(\epsilon_m B - I)x\|^2$ arbitrarily close below $g_m^2(B)$. The supremum of such x shows that the left side of (1.3.1) attains the right side.

EXAMPLE 1.3.2. Consider the example

$$B = \begin{bmatrix} 9 & 0 \\ 0 & 16 \end{bmatrix}.$$

Then the curve $\|\epsilon B - I\|^2$ consists of two parabolic branches, $(9\epsilon - 1)^2$ to the left of $\epsilon_m = 2/25 = 0.08$, and $(16\epsilon - 1)^2$ to the right of ϵ_m. The minimum $g_m^2(B)$ is the intersection of these two branches and is $(7/25)^2 = (0.28)^2 = 0.0784$. We may take

$$x_1 = \begin{pmatrix} 1 \\ 0 \end{pmatrix} \quad \text{and} \quad x_2 = \begin{pmatrix} 0 \\ 1 \end{pmatrix}$$

for $x = \xi x_1 + \eta x_2$ in (1.3.7). Then $\|(\epsilon B - I)x_1\|^2 = (9\epsilon - 1)^2$ and $\|(\epsilon B - I)x_2\|^2 = (16\epsilon - 1)^2$ attain exactly the left and right branches of $\|(\epsilon B - I)\|^2$, respectively. The minima of $\|(\epsilon B - I)x_1\|^2$ and $\|(\epsilon B - I)x_2\|^2$ are zero and are attained at $\epsilon_1 = 1/9 = 0.1111 \cdots$ and $\epsilon_2 = 1/16 = 0.0625$.

The construction of x as given in the proof above proceeds as follows. Equations (1.3.7) and (1.3.8) become the conic system

(1.3.7') $$\xi^2 + \eta^2 = 1$$

(1.3.8') $$a\xi^2 + c\eta^2 = 0$$

where

$$a = 9(1 - 9\epsilon_m)$$

$$c = 16(1 - 16\epsilon_m)$$

From $(1.3.8)'$ we have

$$\xi = \pm \left(-\frac{c}{a}\right)^{1/2} \eta$$

$$= \pm \left(-\frac{16(1 - 16 \cdot 2/25)}{9(1 - 9 \cdot 2/25)}\right)^{1/2} \eta$$

$$= \pm \left(\frac{16}{9}\right)^{1/2} \eta$$

which substituted into $(1.3.7)'$ yields

$$\eta = \pm \left(1 - \frac{c}{a}\right)^{-1/2}$$

$$= \pm \left(1 + \frac{16}{9}\right)^{-1/2}$$

$$= \pm \frac{3}{5}$$

Thus the desired vector $x = \xi x_1 + \eta x_2$ is

(1.3.10)
$$x = \left(\pm \frac{4}{5}, \frac{3}{5}\right)$$

EXAMPLE 1.3.3. For the matrix $B = \begin{bmatrix} m & 0 \\ 0 & M \end{bmatrix}$, $0 < m < M$, the construction of Example 1.3.2 generalizes to

$$-\frac{c}{a} = -\frac{M}{m} \cdot \frac{\left(1 - M\left(\frac{2}{m+M}\right)\right)}{\left(1 - m\left(\frac{2}{m+M}\right)\right)} = \frac{M}{m}$$

and

$$1 - \frac{c}{a} = 1 + \frac{M}{m} = \frac{M + m}{m}$$

so that

(1.3.11)
$$\eta = \pm \frac{m^{1/2}}{(M + m)^{1/2}}$$
$$\xi = \pm \frac{M^{1/2}}{(M + m)^{1/2}}$$

REMARK 1.3.4. The minmax Theorem 1.3.1 was announced in Gustafson [5] and the proof was presented at the Los Angeles Symposium on Inequalities in 1969. The essentials of the proof given here are in those Proceedings, Gustafson [7]. At the Los Angeles Symposium, the author discussed with E. Asplund the question of the extent to which the minmax result would extend to Banach spaces. In Asplund and Ptak [1], this question was answered in the negative, and in more generality. The proof of the minmax theorem given by Asplund and Ptak [1] is different from that given here.

First they assume a smooth minimum to the curve $\|A + \lambda B\|^2$ and argue by compactness that the curve's infimum is the same as that of one of the $\|(A + \lambda B)x\|^2$ family. In the second possibility of an assumed corner minimum, they convert the problem to the special case $\|\epsilon B - I\|$, as in the considerations here, and use the fact that this curve dips below one to reach a contradiction.

However, the arguments of Asplund and Ptak [1] give no constructive information. On the other hand, the original (1968) proof, which we have given here, is constructive. In particular, in Examples 1.3.2 and 1.3.3 above, this proof yields the exact antieigenvectors of the matrix. Antieigenvectors will be discussed in the next section. Generally, the x vectors of the proof are in fact either exact or approximate antieigenvectors. This offers up the possibility that antieigenvectors may have a 'two-component' nature for arbitrary strongly accretive bounded operators B, a result known thus far for normal operators. See Remark 2.1.2.

Chapter 2   Antieigenvalues

2.1. Variational Formulation. In 1969 in Gustafson [7] the terminology *antieigenvalue* was introduced for $\cos A$. More generally, the nth antieigenvalue was defined variationally by

$$(2.1.1) \qquad \mu_n(A) = \inf_{\substack{x \in D(A) \\ x \perp \{x_1,\ldots,x_{n-1}\}}} \frac{\operatorname{Re}\langle Ax, x\rangle}{\|Ax\|\|x\|}$$

where the x_k were called the corresponding *antieigenvectors* of A. It was proposed to consider this as a spectral theory analogous to the usual spectral theory of eigenvalues and eigenvectors.

Total antieigenvalues and corresponding total antieigenvectors were also defined by

$$(2.1.2) \qquad |\mu_1|(A) = \inf_{\substack{x \in D(A) \\ Ax \neq 0}} \frac{|\langle Ax, x\rangle|}{\|Ax\|\|x\|}$$

with the higher ones being defined analogously to (2.1.1). When one thinks of eigenvalues and their corresponding eigenvectors, the latter are those vectors which A dilates but does not turn at all. The eigenvalues are the amount of dilation. The names antieigenvalues and antieigenvectors were chosen to connote the opposite: those vectors which are the critical turnings of A. The corresponding cosines are the antieigenvalues. One could as well identify instead the corresponding angles and call them the anglevalues.

EXAMPLE 2.1.1. Let

$$A = \begin{bmatrix} 9 & 0 \\ 0 & 16 \end{bmatrix}$$

Then $m = \lambda_1 = 9$, $M = \lambda_2 = 16$, $\mu_1 = \cos A = 2\sqrt{2}/3 = 0.9428090416$, $\sin A = 1/3 = 0.3333333333$. By the analysis of Gustafson and Seddighin [1] we know that all (normalized) antieigenvectors of normal matrices are of the form

$$(2.1.3) \qquad x = \left(\frac{\pm\sqrt{\lambda_j}}{\sqrt{\lambda_i + \lambda_j}}, \frac{\sqrt{\lambda_i}}{\sqrt{\lambda_i + \lambda_j}} \right)$$

where i and j index all eigenvalues. In this example one quickly sees that a first antieigenvector is

$$(2.1.4) \qquad x_1 = \left(-\frac{4}{5}, \frac{3}{5} \right)$$

It is easily checked that it is turned by A the maximal amount $\phi(A) = 16.26020471°$.

If one takes second antieigenvectors to be orthogonal to the first antieigenvectors, then necessarily

$$(2.1.5) \qquad x_2 = \left(\frac{3}{5}, \frac{4}{5} \right)$$

However, if we do not require orthogonality, a second first antieigenvector is

$$(2.1.6) \qquad x_1' = \left(\frac{4}{5}, \ \frac{3}{5} \right)$$

Thus we see that generally antieigenvectors will come in pairs. Members of their linear span are generally not antieigenvectors so one should not think in terms of antieigenspaces. And it turns out that orthogonality to delineate higher antieigenvectors is not the most appropriate concept. A more appropriate conceptual view will be described in the next chapter.

REMARK 2.1.2. The proof, Gustafson and Seddighin [1], [2], that antieigenvectors of normal operators are, if one is in the natural diagonal basis of the operator, always vectors $z = (z_1, \ldots, z_n)$ with all components zero except for two and that those two satisfy

$$(2.1.7) \qquad |z_i|^2 = \frac{|\lambda_j|}{|\lambda_j| + |\lambda_i|}, \quad |z_j|^2 = \frac{|\lambda_i|}{|\lambda_j| + |\lambda_i|},$$

utilized Lagrange multiplier techniques. It is interesting therefore to note that the pair of first antieigenvectors found in Example 2.1.1 are the same as the desired vector x in (1.3.11) following from the constructive proof of the MinMax Theorem 1.3.1. Thus we see that the MinMax theorem itself in principle, through its constructive proof, may offer a way to find antieigenvectors, or approximations to them, for arbitrary operators. Generally, the structure of antieigenvectors is not known beyond normal operators.

2.2. The Euler Equation. Let us turn back to the first antieigenvalue μ_1, and derive a fundamental result for it and its corresponding antieigenvector pair. For simplicity we consider only the case A bounded strongly accretive. Consider the antieigenvalue functional

$$\mu(u) = \mathrm{Re} \, \frac{\langle Au, u \rangle}{\|Au\| \|u\|}.$$

THEOREM 2.2.1. The Euler equation for the antieigenvalue functional $\mu(u)$ is

$$(2.2.1) \qquad 2\|Au\|^2 \|u\|^2 (\mathrm{Re}\, A)u - \|u\|^2 \mathrm{Re}\, \langle Au, u \rangle A^* Au - \|Au\|^2 \mathrm{Re}\, \langle Au, u \rangle u = 0.$$

Scalar multiples of solutions are solutions, but the solution space is generally not a subspace. When A is normal, the Euler equation is satisfied not only by the first antieigenvectors but also by all of the eigenvectors of A.

Proof. To find the Euler equation we consider the quantity

$$\frac{d\mu}{dw}\bigg|_{\epsilon=0} = \lim_{\epsilon\to 0} \frac{\frac{\mathrm{Re}\,\langle A(u+\epsilon w),u+\epsilon w\rangle}{\langle A(u+\epsilon w),A(u+\epsilon w)\rangle^{1/2}\langle u+\epsilon w,u+\epsilon w\rangle^{1/2}} - \frac{\mathrm{Re}\,\langle Au,u\rangle}{\langle Au,Au\rangle^{1/2}\langle u,u\rangle^{1/2}}}{\epsilon}.$$

Let the expression on the right hand side be denoted $R_A(u,w,\epsilon)$. We have

$$\epsilon R_A(u,w,\epsilon) = [\mathrm{Re}\,\langle Au,u\rangle + 2\epsilon\mathrm{Re}\,\langle(\mathrm{Re}\,A)u,w\rangle$$

$$+ \epsilon^2\langle Aw,w\rangle]\langle Au,Au\rangle^{1/2}\langle u,u\rangle^{1/2} \div D - [\langle Au,Au\rangle + 2\epsilon\mathrm{Re}\,\langle Au,Aw\rangle$$

$$+ \epsilon^2\langle Aw,Aw\rangle]^{1/2}[\langle u,u\rangle + 2\epsilon\mathrm{Re}\,\langle u,w\rangle$$

$$+ \epsilon^2\langle w,w\rangle]^{1/2}\mathrm{Re}\,\langle Au,u\rangle \div D$$

where D is the common denominator

$$D = \langle A(u+\epsilon w),A(u+\epsilon w)\rangle^{1/2}\langle u+\epsilon w,u+\epsilon w\rangle^{1/2}\langle Au,Au\rangle^{1/2}\langle u,u\rangle^{1/2}.$$

At this point, in deriving the Euler equations for eigenvalues of a selfadjoint operator, one gets a fortuitous cancellation of the ϵ-independent terms in the expression analogous to $\epsilon R_A(u,w,\epsilon)$, and the Euler equation immediately follows. Although that fortuitous situation that does not occur here, we may attempt to mimic it by expanding the two square root bracket expressions of the second numerator term

$$[\langle Au,Au\rangle + x(\epsilon)]61/2 = \langle Au,Au\rangle^{1/2} + \frac{1}{2}\langle Au,Au\rangle^{-1/2}x(\epsilon) - \frac{1}{8}\langle Au,Au\rangle^{-3/2}x^2(\epsilon) + \cdots$$

and

$$[\langle u,u\rangle + y(\epsilon)]^{1/2} = \langle u,u\rangle^{1/2} + \frac{1}{2}\langle u,u\rangle^{-1/2}y(\epsilon) - \frac{1}{8}\langle u,u\rangle^{-3/2}y^2(\epsilon) + \cdots$$

where $x(\epsilon)$ and $y(\epsilon)$ are the ϵ dependent terms, respectively, and where ϵ is sufficiently small relative to $\langle Au,Au\rangle$ and $\langle u,u\rangle$, respectively. Then we obtain $\langle Au,Au\rangle^{1/2}\langle u,u\rangle^{1/2}\mathrm{Re}\,\langle Au,u\rangle$ term cancellations, from which

$$\epsilon R_A(u,w,\epsilon) =$$

$$\frac{[2\epsilon\mathrm{Re}\langle(\mathrm{Re}A)u,w\rangle + \epsilon^2\langle Aw,w\rangle]\langle Au,Au\rangle^{1/2}\langle u,u\rangle^{1/2} - \mathrm{Re}\langle Au,u\rangle[\langle u,u\rangle^{1/2}r(\epsilon) + \langle Au,Au\rangle^{1/2}s(\epsilon)]}{D}$$

where $r(\epsilon)$ and $s(\epsilon)$ denote the remainder terms in the square root series expansions above, to be specific

$$r(\epsilon) = \frac{1}{2}\,\langle Au,Au\rangle^{-1/2}x(\epsilon) - \frac{1}{8}\,\langle Au,Au\rangle^{-3/2}x^2(\epsilon) + \cdots$$

$$s(\epsilon) = \frac{1}{2}\,\langle u,u\rangle^{-1/2}y(\epsilon) - \frac{1}{8}\,\langle u,u\rangle^{-3/2}y^2(\epsilon) + \cdots$$

where

$$x(\epsilon) = 2\epsilon \mathrm{Re}\,\langle Au, Aw \rangle + \epsilon^2 \langle Aw, Aw \rangle$$

$$y(\epsilon) = 2\epsilon \mathrm{Re}\,\langle u, w \rangle + \epsilon^2 \langle w, w \rangle.$$

We may now divide by ϵ, from which

$$D \cdot R_A(u, w, \epsilon) = [2\mathrm{Re}\,\langle (\mathrm{Re}A)u, w \rangle + \epsilon \langle Aw, w \rangle] \| Au \| \| u \|$$

$$- \mathrm{Re}\,\langle Au, u \rangle [\| Au \|^{-1} \| u \| \mathrm{Re}\,\langle Au, Aw \rangle + O(\epsilon)]$$

$$- \mathrm{Re}\,\langle Au, u \rangle [\| u \|^{-1} \| Au \| \mathrm{Re}\,\langle u, w \rangle + O(\epsilon)].$$

Note also that by the above expansions

$$D = [\| Au \| + O(\epsilon)][\| u \| + O(\epsilon)]\| Au \| \| u \| \to \| Au \|^2 \| u \|^2 \text{ as } \epsilon \to 0.$$

Thus in the $\epsilon \to 0$ limit of $R_A(u, w, \epsilon)$, we arrive at

$$\left. \frac{d\mu}{dw} \right|_{\epsilon=0} =$$

$$\frac{2\mathrm{Re}\,\langle (\mathrm{Re}\,A)u, w \rangle \| Au \|^2 \| u \|^2 - \mathrm{Re}\,\langle Au, u \rangle [\| u \|^2 \mathrm{Re}\,\langle Au, Aw \rangle + \| Au \|^2 \mathrm{Re}\,\langle u, w \rangle]}{\| Au \|^3 \| u \|^3}$$

Setting this expression to zero yields

$$2\| Au \|^2 \| u \|^2 \mathrm{Re}\langle (\mathrm{Re}A)u, w \rangle - \mathrm{Re}\langle Au, u \rangle [\| u \|^2 \mathrm{Re}\langle A^*Au, w \rangle + \| Au \|^2 \mathrm{Re}\langle u, w \rangle] = 0$$

for arbitrary w, and hence the Euler equation

$$2\| Au \|^2 \| u \|^2 (\mathrm{Re}\,A)u - \| u \|^2 \mathrm{Re}\langle Au \rangle A^*Au - \| Au \|^2 \mathrm{Re}\langle Au, u \rangle u = 0.$$

As the Euler equation is homogeneous (of order 5), scalar multiples of solutions are solutions. This was immediate from the variational quotient for μ, but as it turns out the implications are greater from the Euler equation (2.2.1): for selfadjoint or normal operators, all eigenvectors also satisfy it. The details of the verification are left to the reader. The following example illustrates how this occurs.

EXAMPLE 2.2.2. Let

$$A = \begin{bmatrix} 2 & 0 \\ 0 & 1 \end{bmatrix} \quad \text{and} \quad x = (x_1, x_2) = (x_1, \lambda x_1)$$

with λ real. Then $\|x\| = (1 + \lambda^2)^{1/2}|x_1|$, $\langle Ax, x \rangle = (2 + \lambda^2)|x_1|^2$, $\|Ax\| = (4 + \lambda^2)^{1/2}|x_1|$, so the antieigenvalue functional $\mu(x)$ is

$$F(\lambda) = \frac{\text{Re}\,\langle Ax, x \rangle}{\|Ax\| \|x\|} = \frac{(2 + \lambda^2)}{(\lambda^4 + 5\lambda^2 + 4)^{\frac{1}{2}}}.$$

Let us also consider the usual Rayleigh quotient

$$Q(\lambda) = \frac{\langle Ax, x \rangle}{\langle x, x \rangle} = \frac{2 + \lambda^2}{1 + \lambda^2}.$$

Then for large $\lambda \to \infty$, we have $F(\lambda) \to 1$, $Q(\lambda) \to 1 = $ the smaller eigenvalue of A, and $x \to (0, 1)$, the corresponding eigenvector. For small $\lambda \to 0$, we have $F(\lambda) \to 1$, $Q(\lambda) \to 2 = $ the larger eigenvalue of A, and $x \to (1, 0)$, the corresponding eigenvector. As we know from our general theory, $F(\lambda)$ attains its minimum $\mu_1 = \frac{2}{3}\sqrt{2} = 0.9428090416$ at the first antieigenvectors $x = (\pm 1, \sqrt{2})$, i.e., at $\lambda = \pm\sqrt{2}$. Checking this against $F(\lambda)$ above, we have $F(\pm\sqrt{2}) = 4/(4 + 10 + 4)^{1/2} = 0.9428090416$.

Calculating the derivative, we have

$$
\begin{aligned}
(2.2.2) \quad & F'(\lambda) \\
& = \frac{(\lambda^4 + 5\lambda^2 + 4)^{\frac{1}{2}}(2\lambda) - (2 + \lambda)^2(2^{-1})(\lambda^4 + 5\lambda^2 + 4)^{-\frac{1}{2}}(4\lambda^3 + 10\lambda)}{\lambda^4 + 5\lambda^2 + 4} \\
& = \frac{(\lambda^4 + 5\lambda^2 + 4)(4\lambda) - (2 + \lambda)^2(4\lambda^3 + 10\lambda)}{2(\lambda^4 + 5\lambda^2 + 4)^{\frac{3}{2}}} \\
& = \frac{\lambda(\lambda^2 - 2)}{(\lambda^4 + 5\lambda^2 + 4)^{\frac{3}{2}}}.
\end{aligned}
$$

Thus $F'(\lambda) = 0$ at $\lambda = 0$ and at $\lambda = \pm\infty$, corresponding to eigenvectors, and at $\lambda = \pm\sqrt{2}$, corresponding to antieigenvectors.

REMARK 2.2.3. Because both antieigenvectors and eigenvectors satisfy the Euler equation (2.2.1) in the selfadjoint and normal operator cases, this theory constitutes a significant extension of the Rayleigh–Ritz variational theory of eigenvectors.

2.3. Kantorovich–Wielandt Inequalities. In the next chapter we will connect the operator trigonometry to convergence questions in computational linear algebra. It was known in the initial 1966–1969 period (e.g., see Gustafson [3]) that there were connections to matrix condition numbers in numerical analysis. But these connections were not pursued until much later, Gustafson [9]. Then it was discovered that antieigenvalues are intimately related to the Kantorovich error bound for the steepest descent algorithm in optimization theory. This development will be described in the next chapter. Here we wish to expose a new connection, to the Kantorovich–Wielandt inequalities of matrix theory.

First, it should be noted that the early derivation of the expression (1.2.4) for $\cos A$ was obtained by methods entirely independent of the Kantorovich inequality, i.e., entirely by convexity properties of $\|\epsilon B - I\|$, as in the Min-Max Theorem 1.3.1. Thus those early considerations may be seen, in retrospect, as a distinct variation, and operator-theoretic generalization, of the Kantorovich theory for matrices. But let us see how $\cos A$ for positive selfadjoint operators follows directly from the Kantorovich inequality, as was noted in Gustafson [9].

THEOREM 2.3.1. For T a strongly positive $(m > 0)$ selfadjoint operator,

$$(2.3.1) \qquad \cos T = \frac{2\sqrt{mM}}{m + M}.$$

Proof. As just mentioned, in retrospect this is most easily seen by use of the Kantorovich Inequality, which we express here in the form

$$(2.3.2) \qquad \max_{\|y\|=1} \{\langle y, Ty\rangle \langle y, T^{-1}y\rangle\} = \frac{1}{4}\left(\sqrt{\frac{M}{m}} + \sqrt{\frac{m}{M}}\right)^2.$$

First we change the $\cos T$ minimization (1.2.1) to a maximization:

$$\left(\min_x \frac{\langle Tx, x\rangle}{\|Tx\| \cdot \|x\|}\right)^{-2} = \left(\max_x \frac{\|Tx\|\|x\|}{\langle Tx, x\rangle}\right)^2 = \max_x \frac{\|Tx\|^2\|x\|^2}{\langle Tx, x\rangle^2}.$$

In the maximization we then replace x with $\tilde{x} = \langle Tx, x\rangle^{-\frac{1}{2}}x$ so that

$$\max_x \frac{\|Tx\|^2\|x\|^2}{\langle Tx, x\rangle^2} = \max_{\langle Tx, x\rangle=1} \|Tx\|^2\|x\|^2 = \max_{\langle Tx, x\rangle=1} \langle T^{1/2}x, T^{3/2}x\rangle\langle T^{1/2}x, T^{-1/2}x\rangle.$$

Then we let $y = T^{1/2}x$ and require that $\|y\|^2 = \langle Tx, x\rangle = 1$, so that the maximization just written is that of Kantorovich:

$$\max_{\|y\|=1} \langle y, Ty\rangle\langle y, T^{-1}y\rangle = \frac{1}{4}\left(\frac{M}{m} + \frac{m}{M} + 2\right) = \frac{M^2 + m^2 + 2mM}{4mM}.$$

Thus $\cos T$ is the inverse square root of this quantity

$$\cos T = \frac{2\sqrt{mM}}{M + m}.$$

Let us now take a further look at the inequalities of Kantorovich and Wielandt. We will see (Gustafson [15]) the true geometrical meaning of those inequalities. To emphasize this assertion, let us refer directly to the excellent treatment of the Kantorovich and Wielandt inequalities in Horne and Johnson [1]. There the authors do place trigonometric meanings to these inequalities. However, because all such previous geometric interpretations of these inequalities begin from the matrix condition number

$$(2.3.3) \qquad \kappa = \frac{M}{m}$$

and not from the operator angle $\phi(A)$ of our operator trigonometry, apparently the fundamental simpler geometric meaning of Theorem 2.3.4 below was missed.

A number of equivalent forms of the Kantorovich–Wielandt Inequality are stated in Horne and Johnson [1]. Specifically for comparison to our operator trigonometry, one has

THEOREM 2.3.2 (Kantorovich–Wielandt). Let $A \in M_n$ be a given nonsingular matrix with spectral condition number κ, and define the angle θ in the first quadrant by

$$(2.3.4) \qquad \cot(\theta/2) = \kappa.$$

Then

$$(2.3.5) \qquad |\langle Ax, Ay \rangle| \leqq \cos\theta \|Ax\| \|Ay\|$$

for every pair of orthogonal vectors x and y. Moreover, there exists an orthonormal pair of vectors x, y for which equality holds.

Thus the geometrical interpretation of $\theta(A)$ is that of the minimum angle between Ax and Ay as x and y range over all possible orthonormal pairs of vectors. In particular, one has

THEOREM 2.3.3 (Kantorovich–Wielandt). Let $B \in M_n$ be a positive definite matrix with eigenvalues $0 < \lambda_1 \leq \lambda_2 \leq \cdots \leq \lambda_n$. Then

$$(2.3.6) \qquad |x^* B y|^2 \leqq \left(\frac{\lambda_n - \lambda_1}{\lambda_n + \lambda_1}\right)^2 (x^* B x)(y^* B y)$$

for every pair of orthogonal vectors x and y.

In (2.3.6) we recognize the quantity $\sin B = (\lambda_n - \lambda_1)/(\lambda_n + \lambda_1)$ of (1.2.4) for our operator angle $\phi(B)$. From this we may establish

THEOREM 2.3.4 (Gustafson [15]). The condition number angle θ of the Kantorovich–Wielandt theory, defined by $\cot(\theta/2) = \kappa$, and whose geometrical interpretation is that of the minimum angle between Ax and Ay as x and y range over all possible orthonormal pairs of vectors, is precisely related to the operator angle $\phi(A)$ of the operator trigonometry, whose geometrical interpretation is that of the maximum turning angle by A on single vectors x ranging over the whole domain, by

$$(2.3.7) \qquad\qquad \cos \phi(A^2) = \sin \theta$$

Proof. To be given in Gustafson [15].

REMARK 2.3.5. As we mentioned above, it seems that a preoccupation with the conventional condition number κ, influenced the earlier geometric interpretations of the Kantorovich–Wielandt inequalities. Nonetheless, now those interpretations involving all orthonormal pairs of vectors shed additional light on the meaning of the basic operator angle $\phi(A)$.

Chapter 3   Computational Linear Algebra

3.1. Optimization Algorithms. The method of *steepest descent* is one of the basic numerical methods in optimization theory. In steepest descent for the minimization of a function f, a basic algorithm is

$$(3.1.1) \qquad x_{k+1} = x_k - \alpha_k \nabla f(x_k)^T.$$

If we restrict attention to the quadratic case, where

$$f(x) = \frac{\langle x, Ax \rangle}{2} - \langle x, b \rangle$$

where A is a symmetric positive definite matrix with eigenvalues $0 < m = \lambda_1 \leqq \lambda_2 \leqq \cdots \leqq \lambda_n = M$, then the point of minimum x^* solves the linear system

$$Ax^* = b.$$

For the quadratic minimization, i.e., the linear solver problem $Ax = b$, the descent algorithm becomes

$$(3.1.2) \qquad x_{k+1} = x_k - \frac{\|y_k\|^2 y_k}{\langle Ay_k, y_k \rangle}$$

where $y_k = Ax_k - b$ is called the residual error. Letting

$$E_A(x) = \frac{\langle (x - x^*), A(x - x^*) \rangle}{2} = f(x) + \frac{\langle x^*, Ax^* \rangle}{2}$$

measure the error in the iterates, one has the well-known (see Luenberger [1]) Kantorovich error bound

$$E_A(x_{k+1}) \leqq \left(1 - \frac{4\lambda_1 \lambda_n}{(\lambda_n + \lambda_1)^2} \right) E_A(x_k).$$

But in terms of $\lambda_1 = m$ and $\lambda_n = M$ and the operator trigonometry which we are presenting in this Part III, this becomes

$$E_A(x_{k+1}) \leqq (1 - \mu_1^2(A)) E_A(x_k)$$

$$= (1 - \cos^2 A) E_A(x_k)$$

$$= (\sin^2 A) E_A(x_k).$$

Thus the error rate of the method, in the $A^{1/2}$ norm $E_A(x)^{1/2}$, is exactly $\sin A$.

THEOREM 3.1.1 (*Trigonometric Convergence*). In quadratic steepest descent, for any initial point x_0, there holds at every step k

$$(3.1.3) \qquad E_A(x_{k+1}) \leqq (\sin^2 A) E_A(x_k).$$

Proof. The discussion above. See Gustafson [9] for more details.

We may interpret this result geometrically as follows. The first antieigenvalue $\mu_1(A) \equiv \cos A$ measures the maximum turning capability of A. Thus the angle $\phi(A)$ is a fundamental constraint on iterative methods involving A. Steepest descent does a good job but cannot converge faster than the maximum distance from x to Ax, which is represented trigonometrically after normalization by the quantity $\sin A$.

More generally in optimization theory for nonquadratic problems, one uses the Hessian of the objective function to guide descent by quadratic approximation. The smallest and largest eigenvalues m and M of the Hessians at each iteration point then determine the convergence rate of the method. Thus under rather general conditions, e.g., objective function $f(x, \ldots, x_n)$ with continuous second partial derivatives and a local minimum to which you are converging, the objective values $f(x_k)$ converge to the minimum linearly with convergence rate bounded by $\sin H$ of the Hessian.

Steepest descent algorithms are usually slow to converge. By contrast conjugate gradient methods have the advantage of converging, if one ignores roundoff, in N iterations, for an $N \times N$ symmetric positive definite matrix A. Similar to the theory for steepest descent, one knows that in the A inner product error measure

$$E_A(x) = \frac{\langle (x - x^*), A(x - x^*) \rangle}{2}$$

the conjugate gradient error rate is governed by (see Luenberger [1])

$$E_A(x_k) \leqq 4 \left(\frac{\sqrt{\kappa(A)} - 1}{\sqrt{\kappa(A)} + 1} \right)^{2k} E_A(x_0)$$

for any initial guess x_0. Remembering that the condition number $\kappa(A) = M/m$, we may rewrite this as

$$\|x_k - x^*\|_{A^{1/2}} \leqq 2 \left(\frac{M^{1/2} - m^{1/2}}{M^{1/2} + m^{1/2}} \right)^k \|x_0 - x^*\|_{A^{1/2}}$$

from which

THEOREM 3.1.2. For A positive definite symmetric matrix, for any initial guess x_0, the conjugate gradient iterates x_k converge to the solution x^* of

$$Ax = b$$

with error rate bounded by

(3.1.4) $$\|x_k - x^*\|_{A^{1/2}} \leqq 2(\sin(A^{1/2}))^k \|x_0 - x^*\|_{A^{1/2}}.$$

Proof. The above discussion and the fact that the spectrum $\sigma(A^{1/2}) = (\sigma(A))^{1/2}$ by the spectral mapping theorem. Recall that for selfadjoint T one knows that $\sin T = (M_T - m_T)/(M_T + m_T)$.

Let us now look at some numerical computations on simple examples.

EXAMPLE 3.1.3. Let $A = \begin{bmatrix} 1 & 0 \\ 0 & 2 \end{bmatrix}$, for which we know $\cos A = 0.9428090416$ and the angle $\phi(A) = 19.47122063°$. Let us solve $Ax = b$, for $b = \begin{pmatrix} 1 \\ 1 \end{pmatrix}$, first by steepest descent, and then by conjugate gradient. The solution $x^* = \begin{pmatrix} 1 \\ 0.5 \end{pmatrix}$ will be sought from initial guess $x_0 = \begin{pmatrix} 0 \\ 0 \end{pmatrix}$.

Here are the results for steepest descent, to five iterations:

x		$E_A(x_n)$	$\mu_1(x_n)$	$\phi(x_n)$
0	0			
0.6667	0.6667	0.08333	0.94868	18.435
0.8889	0.4444	0.00926	0.94868	18.435
0.963	0.5185	0.00103	0.94653	18.820
0.9877	0.4938	0.00011	0.94868	18.435
0.9959	0.5021	0.00001	0.94842	18.482

Here $\mu_1(x_n) = \langle Ax_n, x_n \rangle / \|Ax_n\| \|x_n\|$ measures the cosine of the angle of each iteration. This leads us to remark that the error $E_A(x_n)$ decreases in very near correspondence to its $\sin A$ trigonometric convergence rate estimate. The angle of each iteration is close to $\phi(A)$. Such convergence behavior has been observed previously in steepest descent, but here ita is seen in terms of the operator trigonometric theory.

Using the conjugate gradient algorithm, we find

x		$E_A(x_n)$	$\mu_1(x_n)$	$\phi(x_n)$
0	0			
0.2941	0.5882	0.25692	0.97619	12.529
1	0.5	$3.5e - 15$	0.94868	18.435

Next we consider a less trivial example.

EXAMPLE 3.1.4. Let

$$A = \begin{bmatrix} 20 & 0 & 0 & 0 \\ 0 & 10 & 0 & 0 \\ 0 & 0 & 2 & 0 \\ 0 & 0 & 0 & 1 \end{bmatrix}$$

From $\sin A = 19/21$ we have the angle $\phi(A) = 64.7912347°$. Let us solve $Ax = b = (1,1,1,1)$, for which the solution $x^* = (-.05, 0.1, 0.5, 1)$ is sought from initial guess $x_0 = (0,0,0,0)$

Steepest descent converges very slowly. To achieve an error of 10^{-6}, 64 iterations were required. The beginning and final values are shown in the following table.

	x			$E_A(x_n)$	$\phi(x_n)$
0	0	0	0		
0.1212	0.1212	0.1212	0.1212	0.5826	42.757
0.0078	0.1043	0.1815	0.1912	0.4464	48.550
0.1030	0.0994	0.2533	0.2824	0.3464	57.141
0.0180	0.0999	0.2929	0.3399	0.2710	47.236
0.0905	0.1000	0.3399	0.4148	0.2133	57.029
...	...	...	...	...	...
0.0501	0.1000	0.5000	0.9985	$1.40e-06$	42.758
0.4994	0.1000	0.5000	0.9987	$1.14e-06$	42.793
0.4994	0.1000	0.5000	0.9987	$9.26e-07$	42.758

Again we see error decreasing like the $\sin^2 A = 0.8186$ trigonometric convergence estimate of Theorem 3.1.1. However, the angle of each iteration is less than $\phi(A)$, as it must also accommodate critical subangle effects presumably due to higher antieigenvectors other than those composed of just the first and last eigenvectors. We return to this point below.

Using the conjugate gradient algorithm, we find

	x			$E_A(x_n)$	$\phi(x_n)$
0	0	0	0		
0.0594	0.0297	0.0059	0.0030	0.7667	13.526
0.0498	0.1063	0.0265	0.0133	0.7112	21.526
0.0500	0.09997	0.5901	0.3062	0.2488	41.676
0.04995	0.1000	0.5000	1.0000	$2.9e-08$	42.745
0.0500	0.1000	0.5000	1.0000	$2.9e-10$	42.758

Again we note the departure from the extreme angle $\phi(A)$ to what we may call a mixed convergence angle.

In Section 2.1 we mentioned that, instead of defining higher antieigenvalues and antieigenvectors by orthogonality constraints, a more appropriate conceptual view would be proposed. The critical subangle effects seen in the above convergence patterns provide that more appropriate view. To elaborate, we show in Table III-1 all 64 iterations of the steepest descent iterations for Example 3.1.4. There $E(x_n)$ denotes the error norm $0.5\langle A(x_n - x^*), (x_n - x^*)\rangle$, $c(x_n)$ the local cosine $\langle Ax_n, x_n\rangle/(\|Ax\| \cdot \|x_n\|)$, $\phi(x_n)$ the local turning angle corresponding to the local cosine, and $s(x_n)$ the local sine. To the right of those data is the history of the x vector approximate solutions as the steepest descent proceeds, in its very slow way.

First let us look again at the $\sin^2 A$ convergence rate bound, to see if it conforms to that in the example above. Picking three ratios at random from the left column of Table III-1, we have

$$\frac{0450}{0558} = 0.80645, \quad \frac{1636}{2012} = 0.81312, \quad \frac{92596}{113888} = 0.81304$$

which is in good agreement with $\sin^2 A = 0.81859$.

Second, looking at the local angles $\phi(x_n)$, after the first three iterations, we find a pattern of alternating strictly monotone decrease

$$57.14, 57.02, \ldots, 43.292, 43.192, \ldots, 42.8016, 42.7932$$

and

$$47.24, 45.49, \ldots, 42.773, 42.770, \ldots, 42.7582, 42.7580$$

The fact that this sequence alternates corresponds to a known phenomenon of residual oscillation in gradient descent convergence. However, both the upper valued and lower valued sequences terminate with values $\sin(\phi_n) = 0.67935$ and $\sin(\phi_{n+1}) = 0.67890$ that are not near $\sin A = 0.90476$. To what may we attribute this?

If we look at the x vectors in Table III-1 we notice that the second and third components are converged rather quickly. Thus the residual error is concentrated in the first and fourth components, and this drives the convergence rate in the first column of Table III-1. On the other hand, the more correct part of the evolving solution is concentrated in the second and third components. If we calculate the pairwise matrix trigonometry partial sines from all pairs of eigenvalues, we have

$$\frac{20 - 1}{20 + 1} = 0.904762, \quad \frac{20 - 2}{20 + 2} = 0.818181, \quad \frac{20 - 10}{20 + 10} = 0.333333,$$

$$\frac{10 - 2}{10 + 2} = 0.666666, \quad \frac{10 - 1}{10 + 1} = 0.818181, \quad \frac{2 - 1}{2 + 1} = 0.333333.$$

The matrix trigonometry partial sine corresponding to the second and third eigenvalues is close to that observed for the $\sin(\phi_n)$.

REMARK 3.1.5. The extent to which this is a general occurrence will be investigated elsewhere. Aspects of its occurrence in computational linear algebra will be presented in the next section. However, in any case, it illustrates the conceptual utility of a combinatorial theory of higher antieigenvalues corresponding to the smaller critical angles obtained from higher antieigenvectors constructed from systematically deleted sets of eigenvectors of the matrix.

3.2. Iterative Solvers for Ax=b. For large linear systems

$$(3.2.1) \qquad Ax = b$$

coming out of engineering and other applications, an extensive class of iterative algorithms has been developed. The goal of this section is to establish the general connection between the trigonometric operator theory and the iterative algorithms such as GCR, GCR(k), GCG, PCG, Orthomin, CGN, GMRES, CGS, BCG, MinRes, Lanczos, Arnoldi, and others with the same basic characteristics, e.g., residual minimization in selected iterative subspaces and CG-like convergence in roughly n-steps for $n \times n$ system $Ax = b$. The exploration of many details, further analysis, algorithm improvements, especially for the cases in which A is neither symmetric nor normal, will have to be worked out elsewhere in the coming years.

Specifically, let us turn now to the papers Greenbaum [1], Eisenstat, Elman, Schultz [1], Van der Sluis and Van der Vorst [1], Van der Vorst and Dekker [1], Nachtigal, Reddy, Trefethen [1], Nachtigal, Reichtel, Trefethen [1]. Our intent here is only to establish new connections of the computational trigonometry to these papers, taken chronologically. This will enable an extension of the observations above about steepest descent and the CG algorithm, and the combinatorial antieigenvalue formulation, to the more advanced algorithms CGN, GCR(k), Orthomin, GMRES, BCG, CGS treated in those papers.

Turning first to Greenbaum [1], an early paper analyzing splitting methods in the context of the CG algorithm, we note that the estimate (2.6) obtained by matrix splittings there can be written trigonometrically as

$$(3.2.2) \qquad \frac{\|e^{(k)}\|_A}{\|e^{(0)}\|_A} \leq 2 \left[\left(\frac{\sqrt{\kappa}-1}{\sqrt{\kappa}+1} \right)^k + \left(\frac{\sqrt{\kappa}+1}{\sqrt{\kappa}-1} \right)^k \right]^{-1} = \frac{2}{(\sin A^{1/2})^k + (\csc A^{1/2})^k}.$$

Also note that the general expression (2.5) there with its factors $\lambda_i/(\lambda_i - \lambda_j)$ the same as those in antieigenvector components (2.1.3), (2.1.7) in Section 2.1 here can thus be interpreted trigonometrically. Finally, note that the Tchebyshev polynomial just above (2.6) which is commonly substituted as a somewhat optimal polynomial choice in these considerations, namely,

$$(3.2.3) \qquad P_k(x) = T_k((2x - \lambda_{\max} - \lambda_{\min})/(\lambda_{\max} - \lambda_{\min}))/T_k((-\lambda_{\max} - \lambda_{\min})/(\lambda_{\max} - \lambda_{\min}))$$

also has trigonometric content. The optimizing constant there is $\sin A$, e.g., the normalizing denominator is $T_k(-(\sin A)^{-1})$.

Turning next to Eisenstat, Elman, Schultz [1], we note that the fundamental estimate (3.3) for residual errors of GCR (estimates depending on this particular estimate occur in many of the other papers dealing with some of the other competing iterative methods)

$$(3.2.4) \qquad \|r_i\|_2 \leqq \min_{q_i \in P_i} \|q_i(A)\|_2 \|r_0\|_2 \leqq \left[1 - \frac{(\lambda_{\min}(\operatorname{Re} A))^2}{\lambda_{\max}(A^T A)} \right]^{1/2} \|r_0\|_2$$

may be interpreted and in principle improved trigonometrically. For example, in the A symmetric positive definite case the quantity in the brackets $[1 - m^2/M^2]$ can be improved to $[1 - \cos^2 A]$ and thus the improved convergence rate $\sin A$. To obtain this, let us choose $g_1(z) = 1 + \alpha z$ as in Eisenstat, Elman, Schultz [1] so that we arrive at

$$(3.2.5) \qquad \min_{\alpha < 0} \|I + \alpha A\| = \sin A$$

by the minmax Theorem 1.3.1. This establishes the improvement just mentioned. How far this may be extended into the class of all positive real matrices A and perhaps beyond remains to be seen.

Turning next to Van der Sluis and Van der Vorst [1], where the issue of convergence speedup of CG and the roles of Ritz vectors is addressed, there are three trigonometric observations that can be made. First, inasmuch as we know that the first antieigenvector, in the A symmetric or normal case, is composed from just the first and last eigenvectors, it is perhaps too strong to state categorically that the convergence of the Ritz vectors play no role in the analysis of CG convergence. While it is certainly true and well known as a general proposition that Rayleigh–Ritz variationally characterized energy values converge remarkably quickly to eigenvalues near the true, even when approximated rather poorly from the eigenvector standpoint, the trigonometric interpretation, that the true antieigenvectors control the operator trigonometry and hence all turning angle dynamics as an algorithm progresses toward the solution x^*, serves to raise to at least some higher emphasis the role of the Ritz vectors as they approximate the eigenvectors in their role as weighted components of the antieigenvectors in that convergence process. Secondly, the notion of 'second condition number' espoused in Van der Sluis and Van der Vorst [1] is consistent with the conceptual view of the previous section that higher antieigenvectors should be thought of as combinatorial selections of eigenvectors. Third, the conclusion (for an isolated highest eigenvalue) that the convergence of Ritz values at the upper end of the spectrum will rarely lead to impressive increases in the convergence rate of CG, due to loss of orthogonality, leads us to suspect, in accordance with combinatorial higher antieigenvector theory, that the correct combinations of eigenvectors, not their orthogonality, is what is essential to a better understanding of the convergence of iterative methods. The true test of this will have to come eventually from the unsymmetric case.

Turning next to Van der Vorst and Decker [1] and to the general question of the design of preconditioners for PCG schemes, we would like to propose that a goal alternate to and in some cases superior to that of seeking a low condition number $\kappa(QA)$ would be to seek a high first antieigenvalue μ. As $\mu = 1$ only for the identity I, the value μ of preconditioned QA is a gauge on how close Q is to A^{-1}. This will be discussed further in the next section. We would also suggest that variations on the Chebyshev method (e.g., Van der Vorst and Decker [1, (3.19)]) could be interpreted and perhaps sharpened trigonometrically, inasmuch as the antieigenvalues can be expected to generally effect sharper versions of field of values bounds.

Turning next to Nachtigal, Reddy, Trefethen [1], the authors there make a forceful case for regarding CGN convergence as governed principally by singular values, GMRES convergence as governed principally by eigenvalues, and CGS convergence as governed principally by pseudo-eigenvalues. We conjecture that all such schemes should be regarded as converging according to, at least in part, antieigenvalues or more to the point, antieigenvectors. The latter are fundamental to an understanding of all turnings of an operator A. The singular values measure only the dilation actions of $A^T A$. The eigenvalues measure only the dilation actions of A. The pseudo-eigenvalues measure essentially growth rates of $(\lambda I - A)^{-1}$. None of these entities measure directly the turning angles of A. All of these dilation actions are combined with the angular actions of A through the antieigenvalues and antieigenvectors of A. Moreover, the latter actions could explain the breakdown of BCG and CGS, for the cosines of an operator iteration sequence tend to zero as, for example, the sequence loses positive definiteness. It would also be interesting to relate the error bounds mentioned in Nachtigal, Reddy, Trefethen [1], and in the other cited papers, many of these going back to the basic bound (3.2.4), to the Euler equation (2.2.1), inasmuch as $\operatorname{Re} A$ and $A^T A$ figure prominently in both.

Turning next to Nachtigal, Reichtel, Trefethen [1], we would propose that we need an 'Arnoldi' method for the estimation of antieigenvalues for nonsymmetric A. That is, the 'Arnoldi' methods estimate eigenvalues of A and as it is probable that the largest and smallest eigenvalues for nonsymmetric A will continue to be important ingredients for the calculation/estimation of the first antieigenvalue (real, total, or imaginary) of A, it would be interesting to devise schemes adapted from these Krylov subspaces for the calculation of antieigenvalues, which to date are known only for A symmetric or normal. Hybrid Arnoldi/GMRES schemes for this would be welcome. Such computational antieigenvalue theory would help 'tighten up' the pseudo-eigenvalue theory which is too loose due to its ties to the field of values, which is too large.

As a final illustration of trigonometric interpretation of iterative methods, let us conclude by going back to the classic book, Varga [1]. In his treatment of the ADI schemes (commuting case) so important to numerical partial differential equations, Varga arrives at the expansion

$$(3.2.6) \qquad \rho\left(\prod_{j=1}^{m} H_j V_j\right) = \max_{1 \leq i \leq n} \prod_{j=1}^{m} \left|\frac{r_j - \sigma_i}{r_j + \sigma_i}\right| \left|\frac{r_j - \tau_i}{r_j + \tau_i}\right| < 1$$

to explain a monotone error decrease of an explicit–implicit iterative scheme, as due to a spectral radius less than one. Here the σ_j and τ_j are eigenvalues of H_j and V_j and we observe that the factors on the right hand side of this bound may be regarded as 'local sines.' ADI and ADI-like schemes remain to this day very important for the numerical solution of partial differential equations in two and three dimensions. The new trigonometric interpretations may lead to a better understanding of the noncommutative convergence theory of such schemes.

REMARK 3.2.1. Similar trigonometric interpretations of the "jagged" residual error estimates of Cullum and Greenbaum [1] are also apparent. We conjecture that such "jaggedness" is in fact essential to the good ultimate convergences of those methods. However the full potential of this conjecture will require a better understanding of operator trigonometry of non-normal operators.

3.3. Preconditioning Numbers. In the preceding section, we asserted that the first antieigenvalue μ may have advantages over the usual condition number κ for the theory and design of preconditioners. Preconditioning has become a key barrier issue for linear solvers $Ax = b$ for very large systems from engineering in which higher resolution and finer detail are desired. Counterposed against such desires are the resultant slower convergence times. A preconditioner must effectively represent the properties of A^{-1} but in a coarse efficient fashion.

In Fig. III-2 we show the convergence of a preconditioned conjugate gradient scheme applied to the Navier equations of elasticity. There an irregular finite element grid which used efficiently the parallel architecture features of a massively parallel computer yielded a linear system of 2760 equations. Convergence to within 10^{-5} of the initial residual error was obtained in less than 100 iterations. See Sobh and Gustafson [1] for further details.

Recently we discovered the work of Kaporin [1,2] and others (e.g., see Kolotilina and Yeremin [1]) in Russia. They use a quantity β, defined for an $n \times n$ matrix A with n positive eigenvalues,

$$(3.3.1) \qquad \beta(A) = \left(\frac{\sum_{i=1}^{n} \lambda_i}{n} \right)^n \Big/ \left(\prod_{i=1}^{n} \lambda_i \right).$$

It is known for example that

$$(3.3.2) \qquad 1 \leqq \beta(A)^{1/n} \leqq \kappa(A) \leqq 4\beta(A).$$

A good exposition of the basic properties of the condition number $\beta(A)$ and its use and potential for providing alternative convergence rate estimates for PCG methods may be found in the recent book Axelsson [1]. In that book β is called the K-condition number of A, but we will stay with the lower case greek symbols κ, μ, and β here.

The connection between the first antieigenvalue $\mu(A)$ and the condition number $\beta(A)$ is immediate: for $n = 2$, we have

$$(3.3.3) \qquad \beta(A)^{-\frac{1}{2}} = \mu(A).$$

Kaporin [1,2] was interested in using $\beta(A)$ for use in finding preconditioning matrices H for the matrix equation $Ax = b$ so that

$$(3.3.4) \qquad HAx = c = Hb$$

is a better conditioned system according to $\beta(HA) \sim 1$. We are proposing using $\mu(A)$ for use in seeking preconditioning matrices Q for the matrix equation $Ax = b$ so that $\mu(Q^{-1}A) \sim 1$. When used in that fashion, we propose to call quantities such as κ, μ and β *Preconditioning Numbers*.

Let us now quickly establish the new trigonometric understandings of the Kaporin condition number $\beta(A)$ as realized through the connection to $\mu(A)$ made here. First, as already seen, in its simplest form with all intermediate eigenvalues stripped out, in trigonometric terms (3.3.3) becomes

$$(3.3.5) \qquad\qquad \beta(A) = 1/\cos^2 A.$$

Thus the full $\beta(A)$, in some averaged sense, is a trigonometry entity. Second, the attempted preconditioning strategies $\beta(HA) \sim 1$ mentioned above are, geometrically from the antieigenvalue point of view, an attempt to find H so that

$$(3.3.6) \qquad\qquad \cos(HA) \sim 1.$$

In other words, we are trying to make HA have maximal turning angle

$$(3.3.7) \qquad\qquad \phi(HA) \sim 0.$$

These geometrical understandings just espoused should be contrasted with the usual preconditioning strategy of trying to make

$$(3.3.8) \qquad\qquad \kappa(HA) \sim 1.$$

A little thought will determine essential practical equivalences between the strategies (3.3.7) and (3.3.8) because we conceive that, as a general metatheorem, we can expect the first antieigenvalue of an operator A to depend strongly on the smallest and largest eigenvalues of A. We know this to be precisely the case for symmetric and normal A, for example. But the conceptual difference between strategies (3.3.7) and (3.3.8) is profound: the latter, conventional strategy is to try to 'undilate' A with H, whereas the former, new strategy is to try to 'untwist' A with H!

The relative strengths of the usual precondition number κ and the β precondition number are treated in Axelsson [1]. For example, the β condition number is advantageous in demonstrating the superlinear convergence often experienced in PCG applications. One may easily check by small-n matrix examples that the relative strengths of the precondition number β and μ as a precondition number will depend on how the intermediate eigenvalues cluster toward the ends of the spectrum or weight the upper or lower halves more.

REMARK 3.3.1. A particular advantage of μ as condition number which will have to be explored elsewhere: it needs only good estimates for λ_1 and λ_n, which are often much more available biproducts of iterative methods than is the whole spectrum needed for β. And in our opinion, its greatest potential advantage: we know precisely its geometrical meaning during the iteration dynamics. On the other hand, how well any of κ, β, and μ serve to design preconditioning matrices, as contrasted to measuring the condition of HA after H has been designed, should be viewed as the main goal. Generally the design of preconditioning matrices H has been governed more by the nature of A^{-1} than by condition numbers.

REMARK 3.3.2. In their use of $\beta(HA)$ to measure a preconditioned system, the Russian school had no trigonometric theory for β. Rather, their motivations go back to log norm convergence estimates and the BFGS update strategies of quasi-Newton iterative algorithms. Thus the connection to μ established here provides a trigonometric meaning to β and to those methods which is new.

REMARK 3.3.3. When designing preconditioning strategies, an important consideration which must now be kept in mind is the nature of current computer architectures. For example, in Sobh and Gustafson [1] we chose simple point Jacobi preconditioning because it was natural to a highly parallel computer, the CM-2. In turn, that computer was natural to the element by element and node by node analysis of the elasticity problem being considered.

Comments and Bibliography

The original semigroup perturbation results of Section 1.1 were treated in a Banach space setting. The work of J. Dorroh [1] motivated Theorem 1.1.3, Gustafson [4], which was generalized in Gustafson and Lumer [1]. Probabilistic versions were given in Gustafson and Sato [1]. The entity $\cos A$ in Section 1.2 was introduced in Gustafson [2]. At about the same time, independently M. Krein [1] and H. Wielandt [1] introduced similar notions; see Gustafson [13] for a historical account. The minmax theorem of Section 1.3 was first given in Gustafson [5]. Further operator trigonometry results may be found in Gustafson and Rao [1]. The interesting fact that $\cos A = 0$ for unbounded accretive operators A in a Hilbert space was shown in Gustafson and Zwahlen [1]. A paper bringing the operator trigonometry up-to-date is Gustafson [10].

About ten years ensued after the notion of antieigenvalues were first introduced in 1969, with little activity. Then the papers by C. Davis [1] and B. Mirman [1] appeared, looking at the question of extending the theory from selfadjoint matrices to normal matrices. This was continued in K. Gustafson and M. Seddighin [1], [2]. A paper bringing the antieigenvalue theory up to date is Gustafson [11]. There the Euler equation Theorem 2.2.1 is revisited: the proof given here is from there. The result of Section 2.3, linking the cotangent condition number angle of the Kantorovich–Wielandt inequalities with the operator angle of the antieigenvalue theory, is new, Gustafson [15], and appears here for the first time.

The connection in Section 3.1 between the Kantorovich error bound of optimization theory and $\sin A$ was first given in Gustafson [9]. The extension to the conjugate gradient algorithm was observed in Gustafson [10]. The beginnings of a combinatorial higher antieigenvalue theory may be found in Gustafson [11]. Example 3.1.4 and Table III-1 are taken from Gustafson [12]. The explorations of the use of operator trigonometry for the analysis and design of iterative algorithms, discussed in Section 3.2, follows Gustafson [13], [14]. The connection to the Kaporin condition number β discussed in Section 3.3 was first presented in Gustafson [14].

$E(x_n)$	$c(x_n)$	$\phi(x_n)$	$s(x_n)$	Step size	Components of x			
0.82500011	NaN	NaN	NaN	0	0	0	0	0
0.58257586	0.73424035	42.756934	0.67888961	0.12121212	0.1212	0.1212	0.1212	0.1212
0.44640702	0.66197258	48.549513	0.74952805	0.07963901	0.007787	0.1043	0.1815	0.1912
0.34638262	0.54256886	57.141316	0.84001133	0.11273779	0.103	0.09945	0.2533	0.2824
0.27096459	0.67898607	47.235538	0.73415115	0.080207817	0.018	0.09989	0.2929	0.3399
0.2133377	0.5442214	57.028527	0.83894163	0.11334635	0.09054	0.1	0.3399	0.4148
0.1689617	0.70108658	45.485754	0.71307616	0.080496207	0.02527	0.1	0.3656	0.4619
0.13444033	0.57444489	54.939233	0.81854326	0.11386013	0.08158	0.1	0.3962	0.5231
0.10742065	0.71407521	44.43254	0.70006899	0.080723159	0.03059	0.1	0.413	0.5616
0.086107597	0.60526431	52.752136	0.79602457	0.11425144	0.07494	0.1	0.4329	0.6117
0.069220148	0.72153568	43.818585	0.69237725	0.080890134	0.03459	0.1	0.4437	0.6431
0.055764433	0.63118702	50.862247	0.77563067	0.11453206	0.06988	0.1	0.4566	0.684
0.045009281	0.72596878	43.450501	0.68772766	0.081006952	0.03767	0.1	0.4636	0.7096
0.03637962	0.65194929	49.311269	0.75826257	0.11472475	0.06596	0.1	0.472	0.7429
0.029440673	0.72869647	43.222775	0.68483681	0.081085719	0.04008	0.1	0.4765	0.7638
0.023846887	0.66838264	48.057641	0.74381761	0.11485314	0.06287	0.1	0.4819	0.7909
0.019331187	0.73042697	43.0778	0.68299081	0.081137605	0.04198	0.1	0.4849	0.8079
0.0156797	0.68138868	47.047745	0.73192177	0.11493693	0.06041	0.1	0.4883	0.8299
0.012724347	0.73155469	42.983112	0.68178277	0.08117114	0.04351	0.1	0.4902	0.8437
0.010329843	0.69171804	46.233739	0.72216767	0.11499092	0.05844	0.1	0.4925	0.8617
0.0083886171	0.73230791	42.919775	0.68097366	0.081192657	0.04474	0.1	0.4937	0.8729
0.0068137865	0.69995546	45.576569	0.71418649	0.1150253	0.05684	0.1	0.4951	0.8876
0.0055357241	0.73282254	42.876458	0.68041982	0.081206366	0.04573	0.1	0.4959	0.8967
0.0044980543	0.70654923	45.04516	0.7076639	0.11504714	0.05556	0.1	0.4969	0.9086
0.0036553624	0.73318189	42.846189	0.68003258	0.081215099	0.04653	0.1	0.4974	0.916
0.0029708212	0.71184403	44.614851	0.70233758	0.11506089	0.05451	0.1	0.498	0.9257
0.0024146696	0.73343801	42.824605	0.67975634	0.081220582	0.04718	0.1	0.4983	0.9317
0.0019627439	0.71610671	44.266028	0.69799081	0.11506969	0.05367	0.1	0.4987	0.9396
0.0015954834	0.7336241	42.808918	0.6795555	0.081223994	0.04771	0.1	0.4989	0.9445
0.0012969904	0.71954542	43.983037	0.69444538	0.11507557	0.05298	0.1	0.4992	0.9509
0.0010543733	0.73376173	42.797313	0.67940689	0.081226081	0.04814	0.1	0.4993	0.9549
0.00085716031	0.72232395	43.753315	0.69155485	0.11507881	0.05242	0.1	0.4995	0.96
0.00069684896	0.73386526	42.788581	0.67929506	0.081227534	0.04849	0.1	0.4995	0.9633
0.00056652853	0.72457194	43.56675	0.68919917	0.11508116	0.05197	0.1	0.4997	0.9675
0.00046058552	0.733944	42.781939	0.67920999	0.081228308	0.04877	0.1	0.4997	0.9702
0.00037445797	0.72639257	43.415183	0.68728003	0.11508248	0.0516	0.1	0.4998	0.9736
0.00030443788	0.73400486	42.776806	0.67914422	0.081228934	0.049	0.1	0.4998	0.9757
0.00024751259	0.7278682	43.292026	0.68571706	0.11508298	0.0513	0.1	0.4999	0.9785
					0.04919	0.1	0.4999	0.9803
0.00020123333	0.7340523	42.772803	0.67909294	0.081229158	0.05106	0.1	0.4999	0.9825
0.00016360718	0.72906494	43.191938	0.68444453	0.11508377	0.04934	0.1	0.4999	0.984
0.00013301735	0.73408949	42.769665	0.67905273	0.081229322	0.05086	0.1	0.4999	0.9850
0.00010814707	0.73003602	43.110586	0.68340867	0.11508455	0.04946	0.1	0.4999	0.987
8.7927139E-05	0.73411906	42.76717	0.67902077	0.081229575	0.0507	0.1	0.5	0.9885
7.1487622E-05	0.73082429	43.044458	0.68256564	0.11508433	0.04956	0.1	0.5	0.9894
5.8121735E-05	0.73414254	42.765188	0.67899538	0.081229843	0.05057	0.1	0.5	0.9906
4.7254678E-05	0.73146445	42.990695	0.68187958	0.11508445	0.04964	0.1	0.5	0.9914
3.8419941E-05	0.73416132	42.763604	0.67897508	0.08122924	0.05046	0.1	0.5	0.9924
3.1236796E-05	0.73198444	42.946985	0.68132135	0.11508466	0.04971	0.1	0.5	0.993
2.5396605E-05	0.73417634	42.762337	0.67895884	0.081230223	0.05038	0.1	0.5	0.9938
2.064846E-05	0.73240674	42.91146	0.68086737	0.1150854	0.04977	0.1	0.5	0.9943
1.6788057E-05	0.73418856	42.761305	0.67894563	0.081229612	0.05031	0.1	0.5	0.995
1.3649395E-05	0.73275	42.882566	0.68049794	2.11508503	0.04981	0.1	0.5	0.9954
1.1097489E-05	0.73419833	42.76048	0.67893506	0.081230201	0.05025	0.1	0.5	0.9959
9.0227422E-06	0.73302889	42.859079	0.68019751	0.11508611	0.04984	0.1	0.5	0.9962
7.3359051E-06	0.73420632	42.759806	0.67892642	0.081230529	0.0502	0.1	0.5	0.9967
5.9643339E-06	0.73325557	42.839982	0.67995314	0.1150844	0.04987	0.1	0.5	0.9969
4.8491925E-06	0.73421264	42.759273	0.67891959	0.08122962	0.05016	0.1	0.5	0.9973.
3.9425645E-06	0.7334398	42.824455	0.67975441	0.11508016	0.0499	0.1	0.5	0.9975
3.2055318E-06	0.73421788	42.758831	0.67891391	0.08123229	0.05013	0.1	0.5	0.9978
2.6062901E-06	0.73358958	42.811828	0.67959276	0.11508168	0.04992	0.1	0.5	0.998
2.1190226E-06	0.73422205	42.758478	0.6789094	0.08123128	0.05011	0.1	0.5	0.9982
1.7228143E-06	0.73371118	42.801575	0.67946148	0.11507907	0.04993	0.1	0.5	0.9984
1.4007719E-06	0.73422557	42.758182	0.6789056	0.081229709	0.05009	0.1	0.5	0.9985
1.138878E-06	0.73381025	42.793221	0.67935449	0.11508497	0.04994	0.1	0.5	0.9987
9.259625E-07	0.73422831	42.75795	0.67890263	0.08123181	0.04994	0.1	0.5	0.9987

Table III-1. Example illustrating combinatorial higher antieigenvalues.
From Gustafson [12].

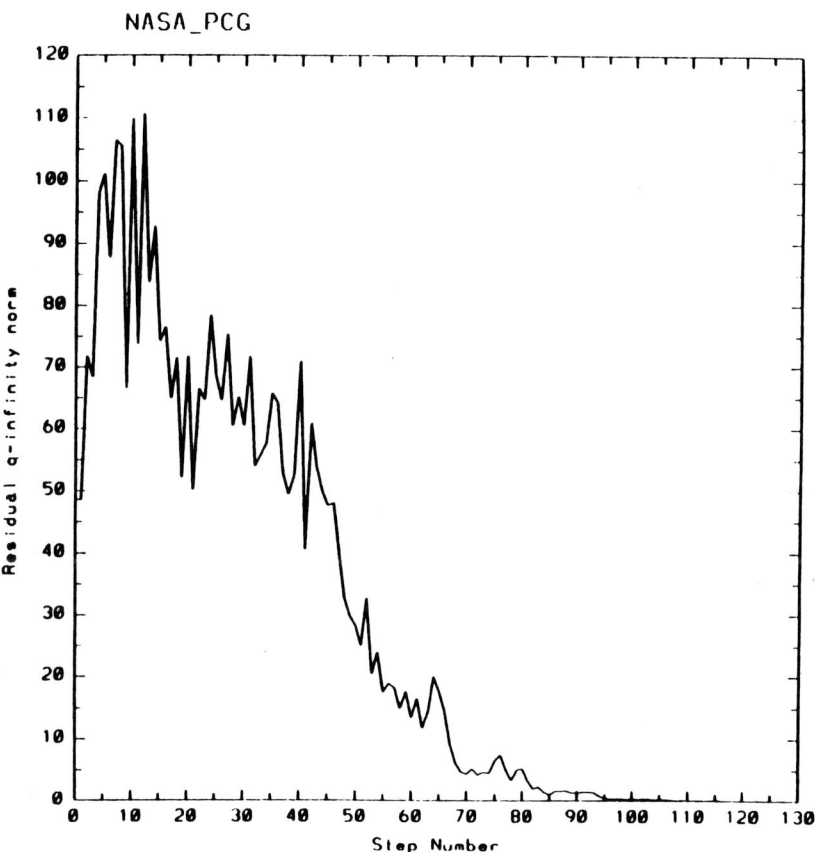

Fig. III-2. Example illustrating preconditioned convergence.
From Sobh and Gustafson [1].

156

References

E. ASPLUND AND V. PTAK

[1] A minimax inequality for operators and a related numerical range, *Acta Math.* **126** (1971), 53–62.

O. AXELSSON

[1] *Iterative Solution Methods*, Cambridge Press, Cambridge (1994).

J. CULLUM AND A. GREENBAUM

[1] Residual relationships within three pairs of iterative algorithms for solving $Ax = b$, to appear.

C. DAVIS

[1] Extending the Kantorovich Inequalities to normal matrices, *Linear Alg. Appl.* **31** (1980), 173–177.

K. DEKKER

See H. A. Van der Vorst.

J. DORROH

[1] Contraction semigroups in a function space, *Pacific J. Math.* **19** (1966), 35–38.

S. EISENSTAT, H. ELMAN AND M. SCHULTZ

[1] Variational iterative methods for nonsymmetric systems of linear equations, *SIAM J. Numer. Anal.* **20** (1983), 345–357.

H. ELMAN

See S. Eisenstat, H. Elman, and M. Schultz.

A. GREENBAUM

[1] Comparison of splittings used with the Conjugate Gradient Algorithm, *Numer. Math.* **33** (1979), 181–194.

See also J. Cullum and A. Greenbaum.

K. GUSTAFSON

[1] A perturbation lemma, *Bull. Amer. Math. Soc.* **72** (1966), 334–338.

[2] The angle of an operator and positive operator products, *Bull. Amer. Math. Soc.* **74** (1968), 488–492.

[3] Positive (noncommuting) operator products and semigroups, *Math. Zeit.* **105** (1968), 160–172.

[4] A note on left multiplication of semigroup generators, *Pacific J. Math.* **24** (1968), 463–465.

[5] A min-max theorem, *Notices Amer. Math Soc.* **15** (1968), 799.

[6] Doubling perturbation sizes and preservation of operator indices in normed linear spaces, *Proc. Cambridge Phil. Soc.* **66** (1969), 281–294.

[7] Antieigenvalue inequalities in operator theory, *Inequalities III*, Los Angeles, 1969, (O. Shisha, ed.), Academic Press, 1972, pp. 115–119.

[8] The RKNG (Rellich, Kato, Nagy, Gustafson) perturbation theory for linear operators in Hilbert and Banach space, *Acta Sci. Math.* **45** (1983), 203–211.

[9] Antieigenvalues in analysis, *Proceedings of the Fourth International Workshop in Analysis and its Applications*, Dubrovnik, Yugoslavia, June 1–10, 1990, (C. Stanojevic and O. Hadzic, eds.), Novi Sad, Yugoslavia (1991), 57–69.

[10] Operator trigonometry, *Linear and Multilinear Algebra* **37** (1994), 139–159.

[11] Antieigenvalues, *Linear Algebra and its Applications* **208/209** (1994), 437–454.

[12] Matrix trigonometry, *Linear Algebra and its Applications*, **217** (1995), 117–140.

[13] Operator angles (Gustafson), matrix singular angles (Wielandt), operator deviations (Krein), *Collected Works of Helmut Wielandt*, Vol. 2, (B. Huppert and H. Schneider, eds.), De Gruyters, Berlin (1996), to appear.

[14] Trigonometric interpretation of iterative methods, *Proc. Conf. Algebraic Multilevel Iteraiton methods with Applications*, Nijmegen, Netherlands, June 13–15, 1996, (O. Axelsson, ed.), 23 –29

[15] The geometrical meaning of the Kantorovich–Wielandt inequalities, to appear.

[16] Article in preparation.

See also N. Sobh and K. Gustafson.

K. GUSTAFSON AND G. LUMER

[1] Multiplicative perturbation of semigroup generators, *Pacific J. Math.* **41** (1972), 731–742.

K. GUSTAFSON AND D. RAO

[1] Numerical range and accretivity of operator products, *J. Math. Anal. Appl.* **60** (1977), 693–702.

K. GUSTAFSON AND K. I. SATO

158

[1] Some perturbation theorems for nonnegative contraction semigroups, *J. Math. Soc. Japan* **21** (1969), 200–204.

K. GUSTAFSON AND M. SEDDIGHIN

[1] Antieigenvalue bounds, *J. Math. Anal. Appl.* **143** (1989), 327–340.

[2] A note on total antieigenvectors, *J. Math. Anal. Appl.* **178** (1993), 603–611.

K. GUSTAFSON AND B. ZWAHLEN

[1] On the cosine of unbounded operators, *Acta Sci. Math.* **30** (1969), 33–34.

R. HORNE AND C. R. JOHNSON

[1] *Matrix Analysis*, Cambridge Press, 1985.

C. R. JOHNSON

See R. Horne and C. R. Johnson.

L. KANTOROVICH

[1] Functional analysis and applied mathematics, *Uspehi Mat. Nauk.* **3**, 6 (1948), 89–185.

I. E. KAPORIN

[1] An alternative approach to estimation of the conjugate gradient iteration number, in *Numerical Methods and Software*, (Y. A. Kuznetsov, ed.), Acad. Sci. USSR, Department of Computational Mathematics (1990), 53–72 (in Russian).

[2] New convergence results and preconditioning strategies for the conjugate gradient method, *Num. Lin. Alg. with Applic.* **1** (1994), 179–210.

T. KATO

[1] *Perturbation Theory for Linear Operators*, 2nd Ed., Springer, New York, 1976.

L. YU KOLOTILINA AND A. YU YEREMIN

[1] Factorized sparse approximate inverse preconditionings I, Theory, *SIAM J. Matrix Anal. Appl.* **14** (1993), 45–58.

M. KREIN

[1] Angular localization of the spectrum of a multiplicative integral in a Hilbert space, *Functional Anal. Appl.* **3** (1969), 89–90.

D. LUENBERGER

[1] *Linear and Nonlinear Programming*, 2nd Edition, Addison–Wesley, 1984.

G. LUMER

See K. Gustafson and G. Lumer.

G. LUMER AND R. PHILLIPS

[1] Dissipative operators in a Banach space, *Pacific J. Math.* **11** (1961), 679–698.

B. MIRMAN

[1] Antieigenvalues: method of estimation and calculation, *Linear Alg. Appl.* **49** (1983), 247–255.

N. NACHTIGAL, S. REDDY, L. TREFETHEN

[1] How fast are nonsymmetric matrix iterations?, *SIAM J. Matrix Anal. Applic.* **13** (1992), 778–795.

N. NACHTIGAL, L. REICHTEL, L. TREFETHEN

[1] A hybrid GMRES algorithm for nonsymmetric linear systems, *SIAM J. Matrix Anal. Applic.* **13** (1992), 796–825.

R. PHILLIPS

See G. Lumer and R. Phillips.

V. PTAK

See E. Asplund and V. Ptak.

D. RAO

See K. Gustafson and D. Rao.

S. REDDY

See N. Nachtigal, S. Reddy, L. Trefethen.

L. REICHTEL

See N. Nachtigal, L. Reichtel, L. Trefethen.

K. I. SATO

See K. Gustafson and K. I. Sato.

M. SCHULTZ

See S. Eisenstat, H. Elman, M. Schulz.

M. SEDDIGHIN

See K. Gustafson and M. Seddighin.

N. SOBH AND K. GUSTAFSON

[1] Preconditioned conjugate gradient and finite element methods for massively data-parallel architectures, *Computer Physics Communications* **65** (1991), 253–267.

160

L. TREFETHEN

See N. Nachtigal, S. Reddy, L. Trefethen;

N. Nachtigal, L. Reichtel, L. Trefethen.

A. VAN DER SLUIS AND H. A. VAN DER VORST

[1] The rate of convergence of conjugate gradients, *Numer. Math.* **48** (1986), 543–560.

H. A. VAN DER VORST

See A. Van der Sluis and H. A. Van der Vorst.

H. A. VAN DER VORST AND K. DEKKER

[1] Conjugate gradient type methods and preconditioning, *J. of Comp. and Appl. Math.* **24** (1988), 73–87.

R. VARGA

[1] *Matrix Iterative Analysis*, Prentice Hall, N.J. (1962).

H. WIELANDT

[1] *Topics in the analytic theory of matrices*, U. of Wisconsin Lecture Notes (1967).

A. YU YEREMIN

See L. Yu Kolotilina and A. Yu Yeremin.

K. YOSIDA

[1] *Functional Analysis*, Springer, New York, 1966.

B. ZWAHLEN

See K. Gustafson and B. Zwahlen.

Comments on the Lectures

A Great Adventurer in Science, and an Endless Subject   163
Takehisa Abe

Comments on Cavity Flow (*with 2 Figures*)   165
Kunio Kuwahara

Brief Comments on PART II. Chapter 1. Probabilistic
and Deterministic Description   167
Izumi Ojima

Professor Karl E. Gustafson and a fragile beauty   169
Mitsuharu Ohtani

A Great Adventurer in Science, and an Endless Subject

Takehisa Abe*

Professor Karl Gustafson. It is hard to introduce Professor Karl Gustafson and all of his work here for me, because it is far beyond my ability. However, I wish to present part of his work being done through the present. His work in mathematical sciences extends from pure and applied mathematics to computational and interdisciplinary sciences. This is a marvelous thing, indeed. To put it briefly concerning the trends in his recent work, there are numerous research papers and talks at international conferences on the following themes:

● *Supercomputers* and *Computational architecture* ●*Wavelets theory, Multidimensional signal analysis* and *Stochastic process* ●*Artificial intelligence* and *Neural networks* ● *Bifurcation theory of nonlinear dynamical systems* and *Theory of vortex motion* ● *Computational fluid dynamics* and *Boundary value problems of nonlinear p.d. equations* (especially, the latter two themes are deeply connected with each other, as shown in Part I) ● *Fractal, chaos* and *Related problems in mathematical physics* and *engineering* (as seen in Part II) ●*Numerical ranges,* and *Antieigenvalues* and *their numerical applications, in operator theory* (the latter is explained in Part III), · · · · ·

And generally these studies are developed closely relating to each other, and naturally spread for various subjects in other fields different from the above, such as *risk problems in financial markets, game theory* and *multivariate analysis.*

He always challenges difficult problems not necessarily come to beautiful results as anyone likes. I see a face as a great scientific adventurer in his research life. His free research activities are supported by an extraordinary curiosity about the unknown and an outstanding physical intuition, in addition to his mathematical genius.

I hope continuations of the above studies including new ones that shall be developed further from now on, and his historical comments on "science and scientists" (e.g. G.Robin and P.Duhem, and the modern science in France). Finally, I would like to dare to say young researchers that those who are afraid of failures in their research or have no interests in other fields different from theirs , should follow at least the example of his adventurous spirit in science, although no everyone can imitate easily him.

Computational Fluid Dynamics. It is well known in the recent fluid dynamics that the structure and the behavior of votices have been clarified by the so-called *vortex methods,* namely numerical approximation methods for the dynamical systems reduced from the Navier-Stokes equations and the vorticity transport equation. Nevertheless, such methods as Lagrangian interpretations of particle trajectories are not generally effective to the case of vortex interactions in the dynamics of vortex motion yet, because of the difficulty of analytically solving "*n*-body problems" (, which are

---

*Faculty of Systems Engineering,
Shibaura Institute of Technology

impossible for $n \geqq 3$ as anyone knows). Probably the methods may still leave more or less room for the overall resolutions to the problems.

On the other hand, we can expect *the general dynamics* of vortex motion as a difficult subject. As to this point , Professor Gustafson proposed "variable *n*-object problems" instead of the above ones (K.Gustafson [3] in Part I). That is, these are problems with nonuniform objects whose individual configurations and number change in time. As he says, the problems would at first sight appear to make themselves even more intractable. However, we know well that their possibility has been already realized in the recent studies, such as replacing point vortices by vortex patches (*Batchelor-Prandtl theory*), contour dynamics (*J.Christiansen, N.Zabusky* and *others*), the random vortex methods (*A.Chorin* and *others*), and the more recent studies (*P.Lax, A.Majda* and *others*) following the evolution of the boundaries of vortex patch equations relating to solutions for the Euler equations.

In the reference above Professor Gustafson has shown and proposed the four new principles of the vortex motions of a viscous incompressible flow based on the examples from cavity flows and airfoil flows. The flow as ocurring in four stages is analyzed in detail : *Generation* of vortex motion → *Evolution* of vortex motion → *Dynamics* of vortex motion → *Limits* of vortex motion. Although the three dimensional case still remains to be solved, the theory is extremely precise and physically refined, and it looks like a sort of axiomatization of the motion. Successively, he discusses the solutions to the boundary value problems of the Navier-Stokes equations and applies these results to the driven cavity poblems (K.Gustafson [4] in Part I). This research gives also the mathematical foundations of the vortex motion mensioned above, and several important remarks apt to be overlooked even by experts. Thus, and as seen in Part I, his discussions are always based on the Navier-Stokes equations themselves, and considers strict and numerically better approximate solutions to boundary value problems of the equations. He does not explicitly accept the above vortex methods. This shows not only importance of the boundary value problems for the Navier-Stokes equations but also that of direct methods capturing real fluid flows. In particular, we expect that the new pressure equation (3.1.10) given by him (, which should be called Gustafson's pressure equation) for the latter purpose will undoubtedly show its usefulness in CFD itself and other applied fields.

It is certainly evaluated that the vortex methods or the dynamical system methods have contributed to depict structures of vortices, and turbulence, that is, a kind of chaos as a geometrical structure. Professor Gustafson is profoundly acquainted with both the concepts, chaos and fractal (as shown in Part II), but he has not referred to turbulence itself as yet. I am sure that he will clarify sooner or later the structure of turbulent flows from the Navier-Stokes equations point of view. Then, we may be able to see a new essential nature of turbulence different from chaotic one. Computational fluid dynamics will develop forever, with theoretical aspects.

Comments on Cavity Flow

Kunio Kuwahara*

Cavity flow is the simplest flow model and it has been widely used to investigate the accuracy, efficiency, etc. of a new scheme. Karl Gustafson studied this problem to check the Moffatt vortices near the corner of the flow and found numerically series of 26 vortices. This is very difficult because the order of the intensity of each vortices is several order of magnitude different. His success of capturing means the numerical method especially the convergence of his iteration is extraordinary good. This computation shows the reliability of the method for very low Reynolds number flows.

However, most of the fluid flows of engineering interest are high Reynolds number flows. Low Reynolds number flow is usually steady and very smooth and 'analytical' because of the viscous diffusion and mathematical treatment is simpler. On the other hand, high Reynolds number flow is usually unsteady and very complicated and not 'analytical'. Therefore mathematical treatment is not well attempted. He showed a Hopf bifurcation of cavity flow of depth 2. This is a first important step to understand a structure of higher Reynolds number flow.

We found even square cavity flow become unsteady at Reynolds number 10000 (see Fig.1). This computation is done by using multi-directional finite difference method with third-order upwinding. The grid is 256*256. At Reynolds number 100000, it becomes chaotic and almost turbulent (see Fig.2). Mathematical justification of high Reynolds number flow computation is strongly wanted.

---

*Institute of Space and Astronautical Science

166

Fig.1 Re=10000, Grid: 256*256

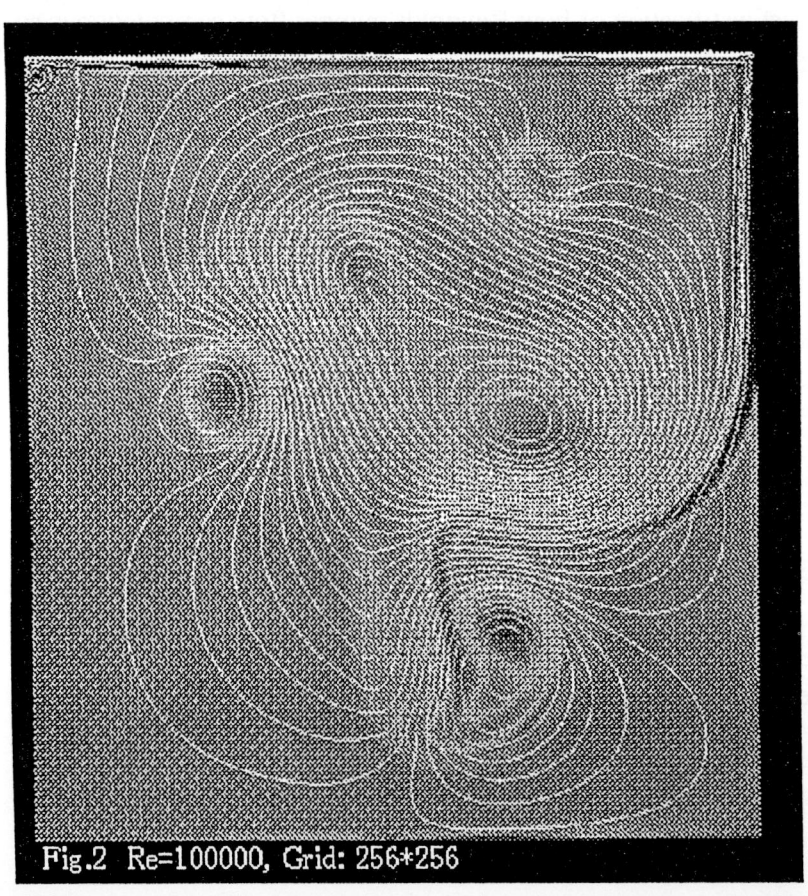

Fig.2 Re=100000, Grid: 256*256

Brief Comments on PART II. Chapter 1.
Probabilistic and Deterministic Description

Izumi Ojima*

Although the essential contents of "1.2 Three Converses" in PART II of Prof.Gustafson's lecture notes consist of rather well known facts, it may be quite useful to have a concise summary of the relevant items in such a coherent form as this. For instance, the first one, Koopman Converse, can be obtained simply by considering the action of $Ad(V_t)^*$ on $Spec$ (=space of pure states=space of characters) of the commutative von Neumann algebra ($L^\infty(\Omega, \mathcal{B}, \mu)$). As for the second and the third ones, one can find similar results (not completely identical, though) in the books, "Harmonic Analysis of Operators on Hilbert Space" by Sz.Nagy and Foias, and "One-parameter Semigroups" by E.B.Davies(to the latter of which I referred the author). As commented in Remark 1.2.10, of course, there are certain important differences in the formulation between the results obtained by Prof.Gustafson with his collaborators and other dilation theories. Here they are viewed as those among the dilations of the Hilbert space with operators in it, of the point dynamics, and of the measure, and the insufficiency of arguments concerning *positivity* in the approaches presented in the above books is pointed out.

It seems, however, that what is more important for understanding properly these differences and their mutual relations is to disentangle the degenerate relations among arguments in L^p-spaces with different p's (at least, with $p = 1, 2, \infty$). Such degeneracy is unavoidable in the discussions restricted to *classical* (i.e. *commutative*) domains, where the similarity between different L^p's are more dominant than their differences. Once we put the problem in the general situation covering not only classical systems but also quantum ones, then the differences become clearer among objects belonging to different L^p's. Namely, we have the following table of correspondences:

quantum	classical
generic observables: C*-algebra $\mathcal{A}$	$\leftrightarrow$ $C(\Omega)$ with $\Omega = Spec(C(\Omega))$
states on $\mathcal{A}$	$\leftrightarrow$ probability measures on Ω
choice of 'standard' state ω	$\leftrightarrow$ 'standard' probability measure μ

* RIMS, Kyoto University

GNS representation $(H_\omega, \pi_\omega, \Omega_\omega)$ $\leftrightarrow$ $(L^2(\Omega, \mu), \pi_\mu, 1)$ with π_μ defined by

$$(\pi_\mu(f)\xi)(\chi) \equiv f(\chi)\xi(\chi) \text{ for } f \in C(\Omega), \xi \in L^2(\Omega, \mu) \text{ and } \chi \in \Omega$$

von Neumann algebra $\pi_\omega(\mathcal{A})''$ $\leftrightarrow$ commutative v.N.alg. $\pi_\mu(C(\Omega))''$

$$= L^\infty(\Omega, \mu)$$

trace-class operators on H_ω $\leftrightarrow$ $L^1(\Omega, \mu)$

(='folium' of state ω or predual $(\pi_\omega(\mathcal{A})'')_*$)

From this table, it should be clear that L^∞, L^1 and L^2 have quite different meaning, and the relation between the latter two is given by

$$L^1 \ni f(\geq 0) \mapsto f^{1/2} \in L^2$$

and

$$L^2 \ni \xi \mapsto |\xi|^2 \in L^1,$$

because of the basic relations given by

$$\mu(A) = \int A(\xi) d\mu(\xi) = \langle 1, \pi_\mu(A)1 \rangle$$

and

$$\mu_f(A) \equiv \int A(\xi)f(\xi)d\mu(\xi) = \int f(\xi)^{1/2}A(\xi)f(\xi)^{1/2}d\mu(\xi) = \langle f^{1/2}, \pi_\mu(A)f^{1/2} \rangle,$$

with $f(\geq 0) \in L^1$. Therefore, the meaning of *positivity* of the maps W_t, W_t^* and Λ need be carefully re-examined. *If* Λ and W_t are to be defined first on $L^2(\Omega, \mathcal{B}, \mu)$ as claimed here, such a relation as $\int_\Omega \Lambda f d\mu = \int_\Omega f d\mu$, for instance, seems to be difficult to maintain, because this is a relation imposed on $f \in L^1$. Likewise, in relation with W_t on L^2, a semigroup dynamics on the observable algebra should take such a form as

$$L^\infty \ni A \mapsto W_t A W_t^* \text{ (or } W_t^* A W_t) \in L^\infty.$$

If $W_t = \Lambda U_t \Lambda^{-1}$, then we have such a relation as

$$W_t A W_t^* = \Lambda[U_t(\Lambda^{-1}A\Lambda^{-1})U_t^{-1}]\Lambda = \Lambda[Ad(U_t)(\Lambda^{-1}A\Lambda^{-1})]\Lambda.$$

In this context, what is to be examined is *not* the positivity of the maps Λ and/or Λ^{-1} on L^2, but that of the maps $A \mapsto \Lambda A\Lambda$ and/or $A \mapsto \Lambda^{-1}A\Lambda^{-1}$ on L^∞. I do not know, however, even whether $\Lambda L^\infty \Lambda \subset L^\infty$ and/or $\Lambda^{-1}L^\infty\Lambda^{-1} \subset L^\infty$ are valid or not.

Editor's Comment : See the author's Remark 1.2.4, where these issues are also mentioned.

Professor Karl E. Gustafson and a fragile beauty

Mitsuharu ÔTANI[†]

It is a great honour for me to be given a chance to write here something related with Professor K. E. Gustafson. Needless to say, the study which he has promoted consists of various fascinating subjects and they form a rich mine. Since his works are too vast for me to appreciate the real value of all works, I am afraid that I am not properly qualified to give a comment on his works. Nevertheless, I can fortunately find that the intersection of his and my interests is not empty. One of major subjects of his works is devoted to the theory of semigroups, especially to the perturbation theory for maximal dissipative operators. The theory of semigroups, established by Hille-Yosida, might be likened to a fragile beauty. In other words, it is a very beautiful and complete abstract theory, but on the other hand it can not cover many important partial differential equations such as the Navier-Stokes equation as its applications. This drawback is caused by the fact that the dissipativity (essentially equivalent to the monotonicity) is very much fragile. For instance, the correspondence : $x \mapsto y = x^3$ is obviously monotone increasing. However, this graph does not remain monotone increasing anymore in the new coordinates (X, Y) generated by the rotation of the (x, y)-coordinates with the angle $\theta > 0$ even though θ is taken very small. This example suggests that even if an operator has the dissipativity in its nature, if the space where it works is not suitably chosen, then it apparently behaves as a non-dissipative operator. In worse situation, it could be the case that there is no such a suitable space where it can work as a dissipative operator. Even in these cases, however, the operator can be regarded as the sum of an appropriate dissipative operator and a variation which can be treated as a (small) perturbation. Here arises the necessity and importance of the study for the theory of the perturbation for maximal dissipative operators. It is Professor Karl E. Gustafson who incarnated this philosophy to set up the perturbation theory for maximal dissipative operators which enlarged extensively the applicability of the theory of semigroups. In this sense, he metamorphosed a fragile beauty into an attractive and able beauty. It is my great pleasure that I could have done some works on the perturbation for nonlinear maximal dissipative operators, which might serve as a nonlinear semigroup version of his theory. I deeply hope that there may come forward many young mathematicians who are inspired by this book and open up new fields in mathematical sciences.

[†]Department of Applied Physics, School of Science and Engineering,
Waseda University, 3-4-1, okubo, shinjuku-ku, Tokyo, Japan